AF488191

Data Engineering with a Focus on Scalable Platforms and Real-Time Analytics

Amarnath Immadisetty

DEDICATION

To my wonderful wife, whose unwavering support and encouragement have been my constant source of strength. Your love and belief in me inspire me every day to reach new heights.

To my children, who remind me daily of the joy in discovery and the power of curiosity. You are my greatest motivation to create a better future for all.

To my parents, whose sacrifices and wisdom laid the foundation for everything I am today. Your values and guidance continue to shape my journey, and I am forever grateful for your love and guidance.

And to the next generation of tech enthusiasts—may you harness the power of innovation, curiosity, and passion to shape a better, more connected world. The digital era is yours to transform, and I hope this work inspires you to embrace the challenges and opportunities that lie ahead.

This work is a reflection of all your support and love. Thank you for being my pillars.

CONTENTS

PREFACE

In the last two decades, data engineering has evolved from the traditional field of data management to the fundamental platform of data-driven organizations. The phenomenon of big data combined with the high rate of innovation has opened a world of opportunities for processing data for real-time, analytics, and business intelligence purposes. However, with these opportunities come some issues: scalability, data security, compliance, and leveraging of advanced technological advancements such as AI and ML.

For the students, practitioners, and fans of data engineering, this book titled Data Engineering with a Focus on Scalable Platforms and Real-Time Analytics offers a one-stop guide. It is intended to deliver fundamental information, working experience, and new trends and tips for data engineering which will help the readers to build effective, strong, and efficient data systems.

The structure of the book enables the reader to advance step by step: from fundamental concepts of data engineering to advanced methods and new trends, as well as applications in various sectors. Areas like building for scale, real-time data processing, and handling security and compliance issues are the subject of modern data platforms and therefore perfectly picked.

Our goal was to provide this book not only as a source of information and techniques but also to kindle imagination, as well as to raise awareness of what data engineering is capable of. May you gain that know-how to deal with real-life issues and come out with significant positive impacts in the ever-evolving field of data engineering, as you read through the chapters.

ACKNOWLEDGEMENT

The process of making this book would not have been a success if not for the cooperation of so many people and companies.

However, I do not know where to start in thanking every single friend and family member who stood by me during the period of writing this paper. This support has motivated me because they always believe in me.

I am grateful to the team of specialists, editors and colleagues, who contributed their expertise, interest and input in making the work technically correct and up to date. I want to use this opportunity to thank you for your contributions which have helped to make this book much richer in terms of content.

Thanks to the publishers for their confidence in this work, and for making it possible; and to the technical team whose professionalism saw to it that this work comes out in the best manner possible.

I am also very thankful to the data engineering community, the practitioners, the researchers, the enthusiasts that have never ceased to inspire me. I am grateful for all the people whose work has informed the content of this book and inspired me to explore this topic.

Finally, I am grateful to the readers of this paper. It is your curiosity and your desire to learn that this book exists. I wish this work equips you to accomplish your objectives and bring improvements to data engineering.

To all those who were with me on this campaign, I want to say thank you very much.

INTRODUCTION TO DATA ENGINEERING

1.1 Chapter Overview

The present chapter aims to give a solid overview of data engineering as an introduction to its key concepts and to highlight its importance for contemporary data environments. Data engineering is concerned with the setup, construction and operation of pathways and frameworks that allow organisations to manage and process big volumes of information. It is crucial nowadays, when data is the king and companies need strategies based on real data to gain benefits and improve their performance. The purpose of the chapter is to describe the development of data engineering approaches based on expanding the initial ETL approach to include real-time data, distributed systems, and cloud solutions. Organizations are struggling to handle this complex data and their challenges arise in the form of data quality, scalability, data security and critical data compliance to regulatory standards. These challenges are even more challenged by the increasing rate of technology and availability of Big Data technologies and the need for qualified workers capable of translating this data into beneficial information. It also analyses the large and evolving set of tools and technologies available to support data engineering. Data ingestion solutions for example Apache Kafka, AWS Kinesis, storage solutions Snowflake, MongoDB, Google Big Query, and processing solutions Apache Spark and Flink are just but a few of the data engineering tools available. Coordination tools such as Apache Airflow and Prefect assist in undertaking data pipelines, and data quality and governance tools are Great Expectations, and Collibra. Tackling these principles, this chapter also underlines the relevance of data engineering for reliably supporting analytical insights as well as decision-makers and creates the backdrop for exploring the current processes, techniques, and developments in the field further. The way that data engineering supports more extensive goals of present-day data-centric organizations is described in detail, thereby, providing readers with a clear vision of what the discipline is about.

1.2 Data Engineering

Preparing data for analytical or operational purposes is the main responsibility of a data engineer, an IT specialist. Among the responsibilities of this profession include creating and constructing systems for data collection, storage, and analysis.

Building data pipelines to combine information from several source systems is usually the responsibility of data engineers. These software developers organize data for use in analytics applications by integrating, combining, and cleaning it. They work to maximize the big data ecosystem within their company and make data easily available.

Each organisation has different data requirements for engineers, especially in terms of size. The analytics infrastructure becomes more complicated and the engineer is tasked with keeping more data as the organisation grows. Some sectors, including healthcare, retail, and financial services, require more data than others.

Together with data science teams, data engineers increase data transparency so that companies can make more reliable business decisions (Gillis et al., 2023).

1. Data engineering process

A number of jobs are included in the data engineering process, which transforms vast amounts of raw data into a useful output that satisfies the requirements of analysts, data scientists, machine learning engineers, and others. The end-to-end process usually includes the following steps.

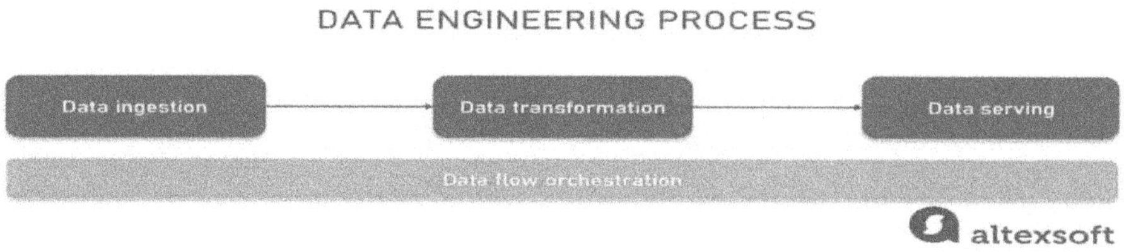

DATA ENGINEERING PROCESS

Figure 1.1: A data engineering process in brief

Source: - *(Science & Engineering, 2023)*

- ***Data ingestion (acquisition)*** transports data from many sources, including websites, streaming services, IoT devices, SQL and NoSQL databases, etc., to a destination system where it is converted for additional analysis. There are many different types of data, including both structured and unstructured data.
- ***Data transformation*** adapts diverse data to end users' requirements. It entails normalizing data, eliminating mistakes and duplication, and formatting the data according to specifications.
- ***Data serving*** provides end users with modified data via a data science team, dashboard, or BI platform.
- ***Data flow orchestration*** ensures that every work is finished by giving insight into the data engineering process. In order to identify and address problems with data quality and performance, it plans and monitors data processes continually.

A data pipeline is the system that automates the data engineering process's intake, transformation, and serving processes (Science & Engineering, 2023).

2. The data engineer's role

The primary responsibility of data engineers is to gather and prepare data for usage by analysts and data scientists. They assume the three primary duties listed below:

Generalists. General-purpose data engineers usually work in small teams, collecting, ingesting, and analyzing data from start to finish. They may be less knowledgeable about systems design yet more skilled than the majority of data engineers. The generalist position would be ideal for a data scientist who aspires to become a data engineer.

The task of developing a dashboard for a small, metro-area food delivery business that shows the quantity of deliveries made daily during the previous month and projects the amount of deliveries for the upcoming month might be taken on by a generalist data engineer.

Pipeline-centric engineers. These data engineers usually work on more complex data science projects across dispersed platforms as part of a data analytics team. Large and midsize businesses are more likely to require this position.

A pipeline-centric project may be undertaken by a local food delivery business to develop a platform that would allow data scientists and analysts to search metadata for delivery details. (Gillis et al., 2023). They may examine the distance travelled and the amount of time spent driving for deliveries during the previous month, then utilize that information in a prediction algorithm to determine the implications for the company's future operations.

Database-centric engineers. These data scientists create, manage, and add data to analytics databases. This position is usually found in larger organisations with several databases. These engineers work with pipelines, optimised databases for effective analysis, and use extract, transform, and load (ETL) techniques to build table schemas. Data is copied into a single target system from several sources using the ETL process.

Designing an analytics database would be a database-centric project at a major national food delivery business. The data engineer would create the database and build the code to transfer the data from the primary application database to the analytics database.

3. Data engineer responsibilities

Data scientists and data engineers frequently collaborate in analytics teams. Data engineers supply data in forms that data scientists may use to execute algorithms and queries on the data for applications such as data mining, machine learning, and predictive analytics. Additionally, data engineers provide analysts, corporate leaders, and other end users with aggregated data so they may evaluate it and use the findings to enhance business operations.

The data that data engineers work with might be either organized or unstructured. Information that can be arranged into a database or other prepared repository is known as structured data. Unstructured data, including text, photos, audio, and video files, does not fit into traditional data models. For data engineers to manage both kinds of data, they need to be aware of the various approaches to data architecture and applications. Additionally, the data engineer's toolset includes a range of big data technologies, including open source data input and processing frameworks.

While each organisation has different specific duties for data engineers, some common duties are as follows:

- Build, test and maintain database pipeline architectures.
- Create methods for data validation.
- Acquire data.
- Clean data.
- Develop data set processes.
- Improve data reliability and quality.
- Develop algorithms to make data usable.
- Prepare data for prescriptive and predictive modeling.

4. Data engineer skill set

Programming languages including C#, Java, Python, R, Ruby, Scala, and SQL are among those that data engineers are proficient in. The three main languages used by data engineers are Python, R, and SQL.

To create and manage data integration jobs, engineers must be well-versed with representational state transfer-oriented APIs and ETL technologies. Additionally, these abilities facilitate easier access to prepared data sets for data analysts and business users.

Data engineers need to know how data lakes and warehouses operate. For example, Hadoop data lakes assist data engineers in their big data analytics efforts by offloading the processing and data storage tasks of traditional business data warehouses.

As NoSQL databases and Apache Spark systems become more prevalent in data processes, data engineers also need to be familiar with them. Additionally, data engineers should be familiar with relational database systems like PostgreSQL and MySQL. Lambda architecture is another area of emphasis since it facilitates unified data pipelines for both batch and real-time processing.

Another key area of attention for data engineers is business intelligence (BI) technologies and their configuration capabilities. They must be able to link to data lakes, warehouses, and other data sources via interactive BI platform dashboards.

To prepare data for machine learning platforms, data engineers need to understand machine learning, even if it is more of a skill set for data scientists or machine learning engineers. They ought to understand how to use machine learning methods and extract knowledge from them.

Finally, it's critical to understand operating systems (OSes) based on Unix. Other operating systems, including Windows and macOS, lack the capability and root access that Unix, Solaris, and Linux offer. For data engineers, they provide more control over the operating system

Companies like IBM and Hadoop vendor Cloudera Inc. have started to provide certifications for data engineers as the field has grown in popularity. The following are a few well-liked data engineer certifications:

- **The Certified Analytics Professional (CAP).** Informs' vendor-neutral certification emphasises data analytics and a candidate's capacity to turn complicated data into insightful knowledge. The test is self-paced, costs money to take, and has prerequisites in education and experience.
- **Cloudera CCP Data Engineer.** The capacity of a candidate to ingest, convert, store, and analyze data in Cloudera's data tool environment is confirmed by this certification. For its four-hour exam, Cloudera charges a fee. Candidates must pass with a minimum score of 70% after completing five to ten practical activities. Although there are no requirements, candidates should have a lot of expertise with data engineering.
- **Google Cloud Professional Data Engineer.** This certification assesses a candidate's proficiency in building and designing data processing systems, ensuring data quality, and using machine learning models. There is a registration fee for the two-hour multiple-choice test offered by Google. Although there are no requirements, Google advises that candidates have some prior knowledge of Google Cloud Platform.
- **AWS Certified Data Engineer - Associate.** The subjects covered in this Amazon Web Services certification include implementing data pipelines, diagnosing problems, and proving proficiency with data-related AWS services. The exam is 130 minutes long and costs money to take (Gillis et al., 2023).

1.3 Importance in the Modern Data Ecosystem

A contemporary data ecosystem consists of a range of services and technology that make it possible to gather, store, process, analyze, and visualise data. This ecosystem helps businesses make data-driven decisions by converting unprocessed data into insightful knowledge.

It is required of a data analyst to be well-versed in the contemporary data ecosystem and how different stakeholders use it. While there has been detailed and complex information and various thoughts around the data ecosystem, here is an attempt to simplify it a bit.

A data ecosystem is a platform that combines raw data from numerous providers and sources, transitioned and processed into a single Enterprise Data Repository and thereby building value through the usage of processed data.

Data from many sources, including photos, videos, streaming data, user discussions, social media platforms, Internet of Things devices, real-time systems, and legacy databases, must be merged into this network of linked, autonomous, and dynamic entities.

This raw data is organized, cleaned and further optimized in a single enterprise data repository. This makes it possible to examine the data, provide insights, and then work with the involved parties to convey and implement the findings(Dasppatnaik, 2024).

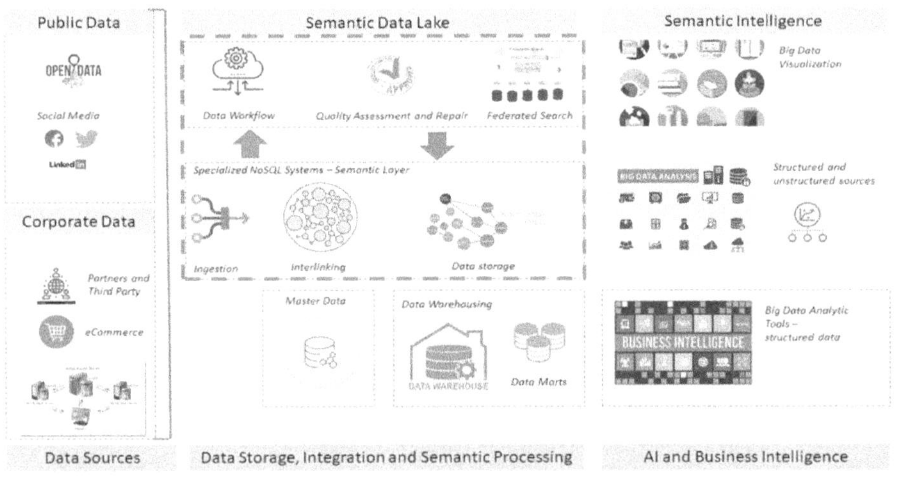

Figure 1.2: Modern data ecosystem

Source: - *(Janev, 2021)*

The modern data ecosystem has five archetypes that have emerged: data utilities, centers of excellence for operations optimisation and efficiency, marketplace platforms, end-to-end cross-sectoral platforms, and ecosystems that prioritise data provision: exchange, availability, and analysis.

The consumption of data from the modern data ecosystem is done by various stakeholders such as Business entities, Products/Apps, Data Analysts, and Data scientists. Each of these stakeholders has a unique role to play in the consumption of data.

- Data is used by business stakeholders to fuel growth and make well-informed decisions. They employ data to find new prospects, streamline procedures, and enhance client satisfaction.
- In order to give their end users individualized experiences, apps employ data to understand user behaviour, preferences, and requirements.
- Analysts use data to generate insights and provide recommendations to business stakeholders. They use data to identify trends, patterns, and anomalies.
- Data scientists use data to build models and algorithms that can be used to make predictions and automate decision-making. They use data to train machine learning models, build predictive & prescriptive models, and develop algorithms (Dasppatnaik, 2024).

Importance of Modern Data Ecosystem

1. Enabling Data-Driven Decision Making

- **Foundation for Analytics:** Data engineering makes it possible to clean raw data into analytically friendly models to help organizations make the right decisions.
- **Support for Business Intelligence:** This forms the foundation of creating the concept of dashboards, KPIs and reports by filling BI tools like Tableau and Power BI with relevant and timely data (Miser & Sarioguz, 2024).

2. Supporting AI and Machine Learning

- **Data Preparation:** For the great success of machine learning, there is a necessity to use high-quality and structurally set data for training purposes. In data engineering, various activities like feature extraction and data pre-processing are also included while they are automated.
- **Scalability for AI Workflows:** Data pipelines deliver the data continually to support real-time or batch training and learning to ensure AI systems remain useful (Sarker, 2021).

3. Handling Big Data Challenges

- **Volume, Variety, Velocity:** n Data Engineering gives an organization the capability of handling more and more data from the sources structured, unstructured or semi-structured in a faster way.
- **Scalable Solutions:** Hadoop, Spark, and the other distributed systems and the cloud, in turn, guarantee that the data structures will scale with the organization (Pappas et al., 2018).

4. Real-Time Decision Making

- **Stream Processing:** In real-time applications like fraud detection, dynamic pricing, and the Internet of Things, real-time analytics may be achieved through the use of technologies like Apache Kafka and Apache Flink.
- **Operational** Intelligence: Helps organization to correspond with live data like the stock market, fluctuations in the supply chain, or interaction with customers (Contribution, 2024).

5. Enhancing Data Quality and Governance

- **Data Consistency:** Data engineering pipelines 'reformatory process to remove inconsistencies and standardization to enhance the quality of the resultant data.
- **Compliance and Security:** Good data management practices are followed when it comes to local and international laws such as

the GDPR and HIPAA have to be complied with to protect the information.

6. Enabling Cross-Functional Teams

- **Seamless Collaboration:** Data engineers create an organizational data layer where people in marketing, finance, operations, and HR can get the data they need in a usable format.
- **Data Democratization:** In structured systems, self-service analytics enable users who may not be so conversant with IT to be able to input, process, and extract information.

<u>Key Elements of Data Ecosystems</u>

Sources of data must be ingested initially. Data scientists then translate, store, and analyze information in an intelligible way prior to the final presentation. The entire procedure is laborious and time-consuming, requiring months to execute.

- **Source data:** Data comes from both internal and external sources. Internal sources are resources that come from within your organisation, such as spreadsheets and proprietary databases. External data sources are those that come from sources outside of your company. Assessing the quality and accuracy of the data sources you use for your project is important.
- **Data storage ETL (Extract, Transform, Load):** The process of getting data ready for analysis is called ETL. It is a catch-all name for a big data ecosystem's data preparation layers. Since there are several types of data, including raw data, organized and unstructured data, etc., managing them effectively typically requires different schemas and alignments.
- **Data warehouses:** The data should be kept in a data lake or warehouse and processed once it has been extracted and converted during the ETL step. This stage of a big data ecosystem is seen by many data science teams as its most crucial element. It's a good idea to keep in mind that data storage in lakes differs from warehouse

storage. The original raw data is preserved in lakes, while data kept in a warehouse is considerably more targeted at the particular analytic activity.

- **Data analysis infrastructure:** The data ecosystem, where all the grunt work is done, depends heavily on analysis. Following collection, ingestion, and preparation, the data is combined and crunched. It is transformed into actionable insights using a number of technologies. Data analysis can be descriptive, predictive, prescriptive, or diagnostic, depending on the specific project.

- **Data visualization:** The visualisation of the data is important. The data should be presented as simple, uncluttered charts to ensure that it is easy to grasp. Users may convert complicated data into understandable charts and graphs with the aid of data visualisation tools. An important first step towards making data-driven, efficient decisions is the deployment of data analytics tools. Among the various data visualisation tools are Looker, Tableau, and Microsoft BI (Koczwara, 2023).

Components of the Modern Data Ecosystem

In the ever-expanding landscape of data engineering, several key components stand out as foundational pillars. These components define how data is handled, processed, and stored within the modern data ecosystem, shaping the capabilities and efficiency of organizations' data-driven initiatives.

- **Batch Processing vs. Stream Processing:** Two essential concepts in data engineering are batch processing and stream processing. Processing massive amounts of data in predetermined groups or batches is known as batch processing. This method works effectively for activities that can tolerate some delay and don't require real-time analysis. Stream processing, on the other hand, processes data in real time, allowing for prompt analysis and reaction. It's perfect for situations like real-time monitoring and fraud detection when prompt insights and action are crucial.

- **Data Warehouses and Data Lakes:** Data warehousing and data lakes are distinct storage architectures designed to accommodate different data types and use cases. Data warehouses are structured repositories optimized for querying and reporting structured data. They are essential to historical analysis and corporate intelligence. In contrast, data lakes do not need the establishment of a schema up front and may contain a wide range of data, including unstructured, semi-structured, and structured data. Data lakes facilitate raw data exploration, machine learning, and advanced analytics.

- **Data Transformation and Orchestration Tools:** Data transformation is the process of transforming unprocessed data into a format that can be used for reporting and analysis. This process encompasses cleaning, enriching, and aggregating data to ensure its quality and relevance. Data orchestration tools, meanwhile, enable the automation and coordination of various data processes and workflows. These solutions ensure that data travels easily and intact from source to destination by scheduling and managing data pipelines (Iabacbdmur, 2023).

1.4 Evolution of Data Engineering Practices

The way that data is seen and utilised has drastically altered over the last ten years. In order to generate innovation and economic value, organisations are increasingly depending on data-driven decision-making and are pursuing and executing data-centric projects. The progression of data engineering from controlling data intake and outflow from databases to increasing accessibility and giving users the appropriate data at the appropriate time will be the main topic of this blog.

The Evolution

A few years ago, as the data environment changed and the need for data handling and engineering assistance expanded, data engineering became more prominent. On the other hand, the discipline was first conceptualized in the late 90s. At the time, data engineering was a subset of a number of emerging data technologies and techniques in the analytics landscape. It dealt with the use of ETL (Extract, Transform and Load) and managed the mobility of data to various channels such as databases and warehouses.

It was in the early 2000s that the requirement of scalability made itself pronounced upon the data landscape as the easy accessibility of the internet brought an increased online engagement between companies and their consumers. This ushered in a new age for data as opportunities for analytics increased dramatically. Organizations began focusing on collecting and storing data through the use of data lakes. Newer technologies such as Apache Hive gained popularity and gave way to increased flexibility and scalability.

In 2006, it became easier and cheaper to store and manage huge amounts of data with the arrival of Big Data and its sibling technologies. Hadoop open-sourcing became a turning point in the way data was being processed. The new complexities behind processing the data gave birth to a new breed of data and backend engineers. Hadoop played an important part in building the storage layer S3 of Amazon Web Services when it was launched in 2006. The second turning point that bridged the gap between these engineers and Big data was when Hive open-sourced in 2010 and this led to a new era of Data engineering (Nineleaps, 2021).

This new era brought forth complexities in data sourcing from different storage locations and organizations were subjected to a new challenge of operating this complex flow of data. This challenge gave rise to data orchestration engines that allowed the flow of data from multiple data storage locations and combined it to make it easily available for various data operations.

This massive explosion of data helped usher in a new age for machine learning. Machine learning models transitioned from being trained on a single machine to being trained on the abundance of data collected from the internet. This further evolved in 2014 with the release of MLlib by Spark for Python which led to the democratization of ML computation on Big Data. In addition to this, Spark offered a new direction for data engineers to compute and process streaming data with relative ease and thus advancing towards the era of real-time processing.

2014 was a significant year in the development of how data can be used. In addition to Spark's impact, the introduction of the Lambda function on Amazon Web Services gave rise to the serverless movement where data ingestion could be easily done without infrastructure management. This allowed data engineers to take a break from managing infrastructure and spend considerably more time on scaling and development. In 2016 the new release, Athena helped propagate things further by allowing to query directly onto s3 without the need to set up a cluster.

The data landscape today has been subjected to a number of iterations and this has changed the way data is now sourced, collected, and processed. In addition to working with data from external sources, organizations of today have to worry about data that is generated internally. The complexity has substantially increased to make sure that the right data is available for experimentation. A Gartner Data Science Team Survey 2018 shows that for data projects, a substantial amount of time goes into tasks such as data collection and preparation, problem analysis, before commencing the development of the various data models. Data engineering has become a critical practice that helps bridge the gaps around the accessibility of data and to ensure success in data and analytics initiatives.

With the gap between newer innovations growing shorter and smaller, the existing layers of data and technology keep shifting towards adaptability. The foreseeable increase in the data volumes from various data sources and collection points will pave the way for newer techniques in reconfiguring data into usable forms. It is becoming vital for organizations to embrace data engineering practice to drive data and analytics success (NINELIPES, 2023).

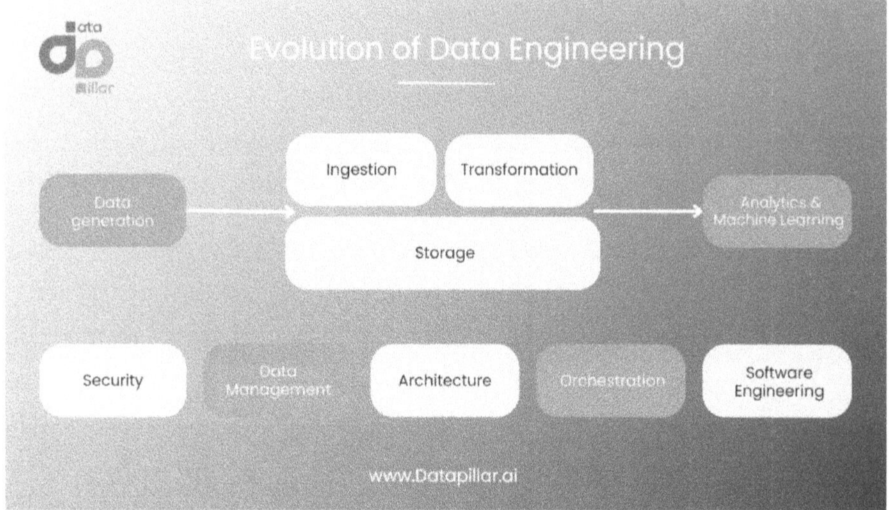

Figure 1.3: Evolution of Data Engineering Practices

Source: - *(frontierinternet, 2022)*

This diagram shows the process of data pipeline starting with data generation, and the processes of ingestion, which is the process of collecting raw data for processing. The data then goes through transformation where it is pre-processed, cleaned, and normalized and is warehoused for readiness use. The gathered data is normally stored then utilized in analytics and machine learning for evaluation and for modelling. Supporting this pipeline are key functions: security enables data protection, data management defines how data must be treated in terms of quality and its lifecycle, architecture defines its design, orchestration handles the running of processes, and software engineering develops the tools and systems that are required. In combination with each other, all these elements make for a strong, fast and secure system of data handling.

The Importance of Data Engineering Over the Decades

The development of technology and business requirements is reflected in the data engineering journey. Data engineers have developed and adjusted to:

1. **Infrastructure Management:** Data engineers have developed methods and systems to efficiently handle data, from cloud platforms to magnetic tapes.
2. **Data Integration:** One of the most important tasks has been to integrate data from many sources, guarantee consistency, and get it ready for analysis.
3. **Performance:** The necessity for speed increased along with the amount of data. Data engineers made sure databases and data processing technologies performed at their best.
4. **Data Quality & Reliability:** Data engineers' role in guaranteeing data correctness, consistency, and dependability became crucial as it was realized that poor data results in poor judgements.
5. **Supporting Advanced Analytics:** These days, data engineering facilitates sophisticated analytics, such as AI and machine learning, laying the groundwork for innovative technical solutions (Krishna, 2023).

Evolution of Data Engineering – The Past

The history of technology and organisational requirements is reflected in the intriguing development of data engineering. Every stage brought new technology and had a big influence on how people and organisations see and use data.

- **1950s-1960s: The Beginnings with File-Based Systems**

Data engineering was primitive in the beginning and focused on file-based systems. Physical formats, including magnetic tapes, punched cards, and paper records, were used to store data. Because human entry and retrieval were necessary for these early systems, data administration was time-consuming. The period was marked by a lack of efficiency and standardisation, with data frequently fragmented and prone to mistakes and duplication.

- **1970s: Advent of Relational Databases**

The creation of relational databases in the 1970s was a significant advancement that drastically altered the field of data engineering. According to Edgar Codd's relational model, data should be stored in tables called relations, with primary and foreign keys preserving the links between the data. Relational database management systems (RDBMS), such as Oracle, were developed as a result of this invention, enabling more effective and organized data retrieval and storage. Data access was further transformed with the introduction of SQL as a common query language, which offered a versatile and potent instrument for data manipulation.

- **1980s: Growth of Personal Computing and Networking**

The 1980s saw a trend towards decentralized data engineering and democratization of data processing with the emergence of personal computers. The advent of PCs with database software made it possible for companies of all sizes to handle data more efficiently. More cooperative and integrated data management techniques resulted from the expansion of local area networks (LANs), which also made it easier for computers to connect and share data. Client-server architectures, which allowed databases to be accessed and maintained remotely and improved operational flexibility and efficiency, also emerged during this time.

- **1990s: Onset of the Internet and Data Warehousing**

The amount of data created and consumed increased dramatically in the 1990s as the internet became widely used. Data warehouses were created as a result of organisations' need for more complex tools to manage the growing volume of data. In order to provide a single, all-inclusive platform for research and reporting, these centralised repositories were made to compile and store data from many sources. In this situation, the ETL procedure became crucial as it made it possible to collect, convert, and load data into the warehouse in a methodical manner from a variety of operational systems. Advanced analytical skills were made possible by data warehousing, which supported corporate intelligence and decision-making procedures.

- **2000s: Big Data and Advanced Analytics**

The big data age, marked by previously unheard-of levels of data volume, diversity, and velocity, began in the twenty-first century. To handle the volume and variety of data, technologies like NoSQL databases and Hadoop, a distributed processing platform, were developed. Scalable and adaptable data engineering solutions that can handle both structured and unstructured data were made possible by these technologies. Businesses now need quick insights to make wise decisions, thus the focus has turned to real-time data processing and analytics. A major advancement in data engineering techniques was the incorporation of advanced analytics, machine learning, and data mining, which made predictive modelling and more sophisticated data-driven tactics possible (Bussa, 2024).

Evolution of Data Engineering – The Present

Data engineering has advanced to a level of complexity and sophistication in the modern period that was nearly unthinkable in its infancy. It is now a fundamental component of the digital economy, supporting the data-driven decision-making processes that are essential for strategic planning, innovation, and corporate operations.

- **Integration of Cloud Technologies**

The extensive use of cloud computing is a characteristic of modern data engineering. Scalable, adaptable, and reasonably priced options for data processing, analysis, and storage are provided by cloud platforms like as AWS, Google Cloud, and Azure. Without requiring a large amount of on-premise technology, they allow data engineers to handle enormous datasets and intricate processing jobs. Additionally, the cloud makes it simpler for international teams to collaborate and share data, removing local restrictions on data access and use.

- **Data Lakes and Real-Time Processing**

One notable development in data management is the emergence of data lakes. Data lakes may store unstructured and semi-structured data, including logs, IoT data, and social media material, without the need for a preset schema, in contrast to data warehouses, which store structured data in an extremely organized fashion. As a result, businesses are able to gather and use more data for analysis and insights. With technologies like Apache Kafka and Spark enabling the streaming and analysis of data as it is created, real-time data processing has also gained significant attention. This allows organisations to get insights instantly and react quickly to changes in the market.

- **Advanced Analytics and Machine Learning Integration**

Data analysis and forecasting capabilities are improved by data engineering's growing integration of machine learning and advanced analytics. In order to facilitate activities like trend analysis, operational optimisation, and consumer behaviour prediction, data engineers collaborate closely with data scientists to design and build data pipelines that feed into machine learning algorithms. Positions like machine learning engineers, which connect data science and data engineering, have emerged as a result of this cooperation.

- **The Democratization of Data**

Finally, data democratization is a defining feature of the current data engineering era. Non-technical users may now undertake complicated data analysis without requiring extensive technical knowledge thanks to self-service analytics tools and platforms. This change promotes a more data-literate culture by enabling more organisational stakeholders to use data in their decision-making processes (Achanta & Bo, 2023).

Evolution of Data Engineering – The Future

Data engineering is expected to continue evolving in the future due to shifting business requirements, increasing data complexity, and technology breakthroughs. The future of data engineering is anticipated to be shaped by a number of significant factors:

- **Artificial Intelligence and Automation**

In data engineering, machine learning (ML) and artificial intelligence (AI) will become increasingly more important in the future. By streamlining data pipelines, AI-powered automation will cut down on mistakes and manual labour. The ability to concentrate on more important projects, such as designing data architectures and decision-support systems, will result from this. Cutting-edge AI algorithms will proactively anticipate and address data problems, improving the dependability and efficiency of data operations.

- **Edge Computing and the Internet of Things (IoT)**

Data engineering will change as a result of the growth of IoT and edge computing, which will concentrate on real-time data processing at network edges. The amount of data created at the edge will increase dramatically as devices get smarter and more connected. To properly handle and analyze enormous data, data engineering must change, requiring more localized and distributed data processing capabilities to enable decisions and actions in real time.

- **Quantum Computing**

Quantum computing is still in its infancy, yet it has the potential to completely transform data processing. Data engineering might be greatly impacted by quantum computing, which has the capacity to handle complicated data sets tenfold faster than conventional computers. This would allow for the study of large data sets in a matter of seconds. Data-driven decision-making would change as a result, enabling almost immediate insights and reactions.

- **Data Privacy and Sovereignty**

Future data engineering techniques must give priority to data security and privacy issues as they continue to escalate. The California Consumer

Privacy Act (CCPA) and regulations such as GDPR are only the first step. Global data sovereignty and privacy laws will likely become stricter, requiring more advanced data governance and compliance systems in data engineering procedures.

- **Collaborative Data Ecosystems**

There will probably be more open and collaborative data ecosystems in the future of data engineering, when data exchange across organisational boundaries becomes standardised and smooth. As a result, new data-driven goods and services will be developed more quickly and with more creativity. The prevalence of data markets and exchanges will increase, enabling moral and effective data sharing among sectors and enterprises(DigitalDefynd, 2024).

1.5 Challenges in Data Engineering

Teams may run into a number of problems as a result of the increasing popularity of data engineering projects and the complexity of use cases. We'll go over the most typical ones below, along with tips on how to handle or avoid them completely. For your convenience, we've divided them into eight categories.

1. Data Ingestion

Data intake is one of the initial issues you may encounter while putting a data engineering method into practice. Since information originates from a number of sources and has a range of forms and structures, the primary problem here is the variety of data. As a result, transformation is necessary before additional processing and analysis.

Real-time data intake, which must be completed quickly, may also provide problems. To manage massive amounts of data and process it instantly, you should consider integrating scalable and effective data input solutions.

Furthermore, we might list quality assurance and data integrity as additional challenges in this section. Inaccurate analysis and insights may result from data that is inconsistent or inaccurate. Therefore, it makes sense to have data validation and cleansing procedures in place to find and fix problems with data quality during intake(Ben Lutkevich, 2023).

Challenges in a nutshell:

- Variety of data sources
- Ensuring data quality and reliability
- Handling large volumes of data
- Real-time data ingestion requirements

2. Data Integration

The second issue we may discuss is data integration, which has to do with how software solutions and data are connected. Integrating data from various systems and efficiently connecting diverse information sources are two of the main objectives of any data engineering project. When working with historical systems that lack the inherent ability to interface with contemporary software, that alone might be difficult.

Find how we integrated a CRM with a VoIP system.

In this sense, updating outdated software should be the first step before stepping up data engineering efforts. By doing this before a project begins, integration issues later on will be reduced.

Data that has to be merged may have different forms, structures, and meanings in addition to diverse systems. To guarantee compatibility and coherence throughout the linked dataset, data transformation, mapping, and schema alignment may be necessary (Sherman, 2015).

Challenges in a nutshell:

- Disparities in the data format and structure
- Working with various technologies and data systems
- Difficulties with data transformation and mapping
- Dealing with concerns related to data governance and compliance

3. Data Storage

The field of data storage faces two major obstacles. The first one is about making room for the growing number of datasets. Data storage systems must have the ability to scale smoothly in order to guarantee that.

Data engineers, for instance, may take use of solutions like cloud-based storage services and distributed file systems, which are easily expandable as data needs increase without sacrificing efficiency or driving up prices.

Organizing and retrieving data is the second problem. It can be challenging to arrange data in a way that enables effective and quick retrieval when vast volumes of data are stored across several platforms.

To maximize data access patterns and reduce retrieval time, effective data indexing, partitioning, and data structure design are essential.

To maximize storage capacity utilisation without compromising data integrity or accessibility, data engineers must also take into account the application of compression and data encoding techniques (Ibrahem & Zeebaree, 2024).

Challenges in a nutshell:

- Selecting appropriate technologies for data storage
- Performance and scalability factors
- Techniques for indexing and data splitting
- Privacy and data security issues

4. Data Processing

Businesses are producing an increasing amount of digital information every day. Data is continuously being produced by IoT devices, smartphone apps, and other platforms. The apparently endless flow of data may quickly overwhelm one.

Conventional processing methods might not be able to effectively manage such high volumes. Data engineers frequently use distributed computing frameworks, like Apache Hadoop or Apache Spark, to overcome this difficulty. These frameworks allow for parallel processing over a cluster of devices, resulting in quicker and more scalable data processing.

The accuracy and validity of analytical results may be impacted by data that is lacking, inaccurate, or inconsistent, which is another problem that might occur in this area. Inaccuracies may arise if many systems are utilising identical digital data without real-time updates. Since low-quality data is useless for your company, it goes without saying that you want to avoid this.

Establishing a thorough data management strategy with a data governance plan is one potential way to address this data engineering difficulty. By doing this, you can make sure that there is someone in control of all data-related operations and that there are regulations in place to assist preserve the accuracy of all of your digital data (Evans, 2023).

Challenges in a nutshell:

- Large-scale data processing
- Large-scale data processing
- Intricate data aggregations and transformations
- Improving pipelines for data processing

5. Processing data at scale

As your business expands, managing your data becomes more and more complicated. To process this data, a model that can grow with these increases is needed.

The proper storage is necessary for efficient data processing.

There will be serious issues with accessing and using your data if you utilize several data storage choices, either alone or in combination, and they are not configured to meet your needs and are unable to manage the format or formats of your data.

To be actionable, you must make sure the data is properly reconciled, and in the process, it must be secure and compatible with the law.

6. Distributed computing and parallel processing

These methods necessitate the use of one or more processors (distributed computing employs numerous separate computers, whereas parallel processing usually requires one computer with multiple processors).

The difficulty of data engineering is figuring out what configuration and operation best suits your needs and making sure that every system—whether it's a single multi-processor device or a standalone machine—is functioning and properly integrated into activities.

7. Complex data transformation and aggregations

Operations will get increasingly sophisticated as you gather and evolve your data.

Managing the process of gathering (aggregating) and transforming your data for various platforms can be difficult.

Establishing a purposeful and dependable procedure to guarantee that the data arrives at its destination in the correct format, free from data corruption or system failures, and managing increasingly complicated activities (sometimes with automation) are made possible by an appropriate data engineering process.

8. Optimising data processing pipelines

As your firm grows, your systems must be able to adapt and handle more data.

If your data pipelines aren't working properly or you're utilising ones that can't handle more datasets, operations won't be as efficient as they should be. Worse, there may be data loss or corruption.

Data pipelines are in charge of getting your data into the appropriate locations at the appropriate times. For you to be able to access and utilize your data without restriction, they must be properly set up (Pratt, 2024).

1.6 Tools and Technologies Landscape in Data Engineering

To extract data from a variety of sources, such as relational databases, NoSQL databases, log files, and other sources, a data engineer may employ a range of techniques and technologies. After being extracted, the data can be changed into a new format and added to a database.

Although the tech world was enthralled with the widespread hype surrounding ChatGPT and Generative AI, 2023 was yet another exciting and dynamic year in the data engineering landscape, with constant innovation and evolution occurring across all levels of the analytical hierarchy and the field steadily growing more diverse and sophisticated.

Data engineers now have many more options because to the ongoing growth of open source tools, frameworks, and solutions! In a world that is changing so quickly, it is crucial to keep up with the newest trends and innovations. In the face of changing data engineering difficulties, the ability to select the appropriate tool for the job is essential for efficiency and relevance.

As a senior data engineer and consultant who has constantly monitored data engineering trends, I would want to share the open source data engineering scene at the start of 2024. an involves highlighting important ongoing initiatives and well-known resources, enabling readers to make wise choices while negotiating an ever-changing technology environment (Mullins, 2022).

Why Present Another Landscape?

Why go through the trouble of presenting another data landscape? Similar periodic studies exist, such as the well-known MAD Landscape, State of Data Engineering, and Repoint Open Source Top 25, but the landscape I'm giving is exclusively about open source technologies that are primarily relevant to data platforms and the data engineering lifecycle.

The MAD Landscape offers a highly thorough overview of all available commercial and open source tools and services for data, AI, and machine learning, whereas the landscape shown here offers a more thorough overview of ongoing open-source initiatives in the Data section of MAD. In contrast to other studies like Repoint Open-Source Top 25 and Data50, which concentrate more on SaaS providers and startups, this study concentrates on open-source projects rather than SaaS services.

Although yearly studies and surveys like Stack Overflow, OSS Insight, and GitHub State of Open Source are excellent resources for learning about community usage and trends, they only cover a small portion of the entire data environment (such as databases and languages).

I have therefore assembled the open source tools and services in the data engineering ecosystem since I am interested in open-source data stacks.

So without further due, here is the 2024 Open Source Data Engineering Ecosystem:

Figure 1.4: OSS Data Engineering Landscape 2024

Source: - (Sadeghi, 2024)

<u>Tool Selection Criteria</u>

Since there are so many open-source projects in each area, it is not feasible to list every tool and service. Consequently, when choosing the tools for each category, I adhered to the following standards:

- Projects that have been retired, archived, or abandoned are not included. Apache Sqoop, Scribe, and Apache Apex are a few well-known discontinued projects that may still be utilised in some commercial settings.

- Projects that are rarely discussed in the community and have not been active on GitHub for a full year are not included. The Apache Pig and Apache Oozie projects are noteworthy examples.

- Projects that are still relatively new and have not received many blog entries, showcases, GitHub stars, forks, or mentions in online communities are not included. Nonetheless, a few encouraging initiatives are highlighted, like One Table, which has gained significant traction and is built on top of current technologies.

- Since I'm primarily interested in things that pertain to the field of data engineering, data science, machine learning, and artificial intelligence tools are not included, with the exception of ML platform and infrastructure tools.

- A list of several storage system types is provided, including relational OLTP and embedded database systems. This is because, even if they are not a component of the analytics stack, the data engineering discipline deals with a wide variety of internal and external storage systems utilised in applications and operations systems (BSS).

- Considering the tool's position in the data stack, the category names are as generic as possible. The primary database model and database workload (OLTP, OLAP) are used to categorise and label storage systems; nevertheless, in the market, "Distributed SQL DBMS" are also known as HTAP or scalable SQL databases.

- Certain tools may fall under more than one category. Volt DB is a distributed SQL DBMS in addition to an in-memory database. However, I have attempted to classify them under the heading that best describes them in the marketplace.

- There might be a hazy distinction between the categories to which some database systems truly fall. For instance, By Comity is based on Click House, a Real-time OLAP engine, yet it makes the claim that it is a data warehousing solution. As a result, it is still uncertain if this OLAP system is real-time (able to support sub-second queries) or not (Sadeghi, 2024).

Top Data Engineering Tools

1. Python

Python has been becoming more and more popular among data engineers because to its adaptability, versatility, and ease of use.

Compared to other languages, Python's built-in libraries make it simple to build code with less lines. It entails spending more time on the real work of being a data engineer and less time writing code!

2. SQL

Structured Query Language is what SQL stands for. Relational databases can be accessed using this language. It is the most extensively used, well-liked, and prevalent language for data management.

3. PostgreSQL

The most dependable, safe, and effective open-source relational database is PostgreSQL. With an emphasis on data integrity, security, and performance, it includes all the tools you need to execute your work.

One of the open-source databases with comprehensive enterprise features, such as advanced authentication, replication, backup/restore, web client libraries, and language APIs, is this one.

4. MongoDB

MongoDB is a free and open-source database that facilitates the development and expansion of cloud applications.

MongoDB maps and indexes data automatically, so you never need to instruct it on how to do it. You can use your preferred programming language to store and query data because it is based on JSON documents. Additionally, because of its extreme speed, you may create apps without worrying about performance snags.

5. Apache Spark

An open-source cluster computing platform called Apache Spark was created to handle large amounts of data. Big businesses and organisations all across the world use it, such as Yahoo!, Netflix, and Spotify.

Spark was created to manage machine learning algorithms as well as batch and stream processing techniques. It can operate independently or in Hadoop clusters.

It is supported by a robust community and significant corporations like Microsoft, IBM, and Intel that have made significant investments in its advancement.

6. Apache Kafka

One tool that can assist you in creating a data pipeline that can manage enormous volumes of data is Apache Kafka. It's widely used by big firms and financial institutions, but it's also a great option for smaller enterprises.

Real-time ingestion and processing of any kind of communication is possible with Kafka. Because of its built-in high-availability characteristics, your data is always accessible when you need it. Messages are stored in topics for convenience.

7. Amazon Redshift

The most potent, scalable, and economical data warehousing solution available today is Amazon Redshift. It is dependable, quick, and simple to use.

All your data from many sources can be analyzed in one location using Amazon Redshift. By processing data on all nodes at once, parallel SQL queries allow you to query hundreds of billions of rows in a matter of seconds. Redshift also automatically manages your clusters, so you don't have to bother about backup and recovery.

8. Snowflake

All of your data can be managed, stored, and analyzed using Snowflake, a cloud data warehouse. Clusters may be automatically created and scaled up or down as needed. Snowflake is an extremely versatile tool for engineers since it supports well-known programming languages like Python and JavaScript.

9. Amazon Athena

Users may use conventional SQL to query data in Amazon S3 using Amazon Athena, a fully-managed data service. It is perfect for ad hoc analysis, interactive queries, and basic visualizations since it is user-friendly and has a strong feature set.

If you want to conduct SQL queries on data stored in Amazon S3 without having to worry about infrastructure management or scaling up as needs change, Athena is the ideal option.

10. Apache Airflow

A program called Apache Airflow was developed to assist you in managing your data pipelines. Data pipeline construction, monitoring, and optimisation are made simple with a workflow scheduler.

Anything that has to be done repeatedly on big datasets can be done with Apache Airflow. It covers tasks including data analysis, machine learning, and ETL. Additionally, more intricate processes than just tasks or scripts (such webhooks) may be created (Raschka et al., 2020).

Advantages of Using Data Engineering Tools

1. **Improved Data Quality:** To guarantee data integrity, correctness, and consistency, data engineering tools apply complex algorithms and procedures. They can improve data quality by automatically identifying and fixing mistakes, eliminating duplication, and validating data to predetermined standards. This enhancement is

essential for producing trustworthy insights and facilitating well-informed decision-making.

2. **Decision-making:** Organisations may use machine learning models and advanced analytics to find trends, patterns, and insights when they have access to high-quality, processed data. Stakeholders can quickly make data-driven choices thanks to data engineering tools, which make it easier to aggregate and visualise data. Competitive advantage, operational effectiveness, and strategic planning can all be greatly impacted by this skill.

3. **Cost Savings:** Through the automation of formerly manual operations, data engineering tools can result in considerable cost reductions. Labour expenses are decreased by automation as it eliminates the need for significant human interaction. These technologies may also optimize processing and storage of data, which reduces infrastructure expenses and enhances resource utility.

4. **Automated ETL Processing:** Data integration from several sources requires the use of ETL (Extract, Transform, Load) procedures. By automating these procedures, data engineering solutions guarantee effective data loading into a centralised repository, transformation, and consolidation. Automation speeds up data availability and reduces mistakes, allowing for fast reporting and analysis.

5. **Organization:** Large volumes of data may be structured and arranged with the use of data engineering technologies. They make it possible to classify data, develop logical data models, and put metadata management procedures into action. This company makes sure that data is useable, clear, and readily available.

6. **Building Data Pipelines:** Designing and overseeing data pipelines that automate data flow from source to destination requires the use of these tools. Data pipelines make it easier for data to flow through different processing steps and guarantee that data is accessible when and where it is needed.

7. **Data Cleansing:** To guarantee the quality of data, data cleaning is an essential procedure. Advanced features for cleaning data by

locating irregularities, discrepancies, and unnecessary information are provided by data engineering tools. Data accuracy and dependability are enhanced by cleaning, which is essential for analytical models and decision-making procedures.

8. **Developing Data Models:** Data models are essential for comprehending and efficiently using data. The creation of intricate data models that accurately depict the entities and connections found in the data is made possible by data engineering tools. These models, which offer an organized framework for data interpretation, are crucial for analytics, reporting, and application development.

9. **Efficiency:** Data engineering technologies greatly improve operational efficiency through automation and optimisation. Instead of wasting time on repetitive data management operations, they facilitate quicker data processing, lower manual mistakes, and free up data experts to concentrate on higher-value activities like analysis and strategy formulation.

10. **Enhancing Information Security:** Strong security measures are included into data engineering tools to shield data from breaches, illegal access, and other online dangers. They include features like auditing, access control, encryption, and compliance that assist protect private data and guarantee that data management procedures follow legal requirements.

11. **Real-time Data Processing:** Real-time data processing and analysis is crucial in the fast-paced world of today. Real-time data intake, processing, and analysis are made easier by data engineering tools, which let businesses react quickly to changing consumer behaviour, operational shifts, and new trends. (Simplilearn, 2024).

1.7 Chapter Conclusion

In this chapter, we argued about what data engineering is and why it is important in the current and developing data landscape. Data engineering plays a crucial role to the functioning of organizations that are doing business in the data-driven environment by providing the necessary infrastructure for processing and transferring data from its source to consumption. Its relevance has increased sharply today, where data is no longer limited to being generated as result of organizational processes, but is now proving to be a key resource for creating value. This chapter gave an overview of history of Data Engineering from basic ETL to advanced concepts used in today's world.

While data engineering unfolds a range of possibilities, it is not an easy journey. Concepts like data quality, scalability, security and governance point to the fact that there is need for strong approaches and better tools. We also analyze a wide variety of technologies that are being used by data engineers across ingestion and storage, orchestration, processing, and governance, which form the core of the data value chain required to build robust data pipelines.

However, it has been noticed that as organizations embark on their journey of digitalization, the need for better, cheaper and faster data engineering practices will also rise. This chapter provides the initial framework for the position of data engineering in the larger conceptual framework of data management and analysis, paving the way for a more detailed look at its approaches, issues and tools in the subsequent chapters. Thus, data engineering that brings meaningful connection between heaps of plain data and valuable information still continues to be one of the critical fundamentals of business and technologization.

Multiple Choice Questions (MCQs)

1. What is the primary role of a data engineer?

    a.   To analyze data trends

    b.   To prepare data for analytical or operational use

    c.   To visualize data

    d.   To design machine learning models

2. Which of the following is NOT a stage in the data engineering process?

   a. Data ingestion
   b. Data transformation
   c. Data visualization
   d. Data serving

3. Data pipelines automate which steps in the data engineering process?

   a. Ingestion, transformation, and serving
   b. Data modeling and prediction
   c. Visualization and reporting
   d. Data cleaning and storage

4. Which database type is used for unstructured data?

   a. SQL databases
   b. NoSQL databases
   c. Relational databases
   d. Analytical databases

5. What does ETL stand for?

   a. Extract, Transform, Load
   b. Evaluate, Test, Learn
   c. Encode, Transfer, List
   d. Export, Tabulate, Load

6. What is the main focus of pipeline-centric data engineers?

   a. Data visualization
   b. Managing distributed systems
   c. Database administration
   d. Writing machine learning algorithms

8. Which programming language is commonly used in data engineering?

a. HTML
b. Python
c. C++
d. Ruby

9. Data governance frameworks ensure:

a. Data accessibility
b. Data security and compliance
c. Faster data processing
d. Visual data representation

10. The "modern data ecosystem" includes:

12. Data generation, analysis, and visualization
13. Storage, streaming, and querying of data
14. Big data technologies only
15. Manual data collection processes
16. **What is the significance of data democratization?**
17. Enables technical experts only to use data
18. Allows non-technical users to access and analyze data
19. Limits access to data for security
20. Prioritizes visual representation over raw data

Answer

1	2	3	4	5	6	7	8	9	10
b	c	a	b	a	b	b	b	a	b

DESIGNING SCALABLE DATA PLATFORMS

2.1 Chapter Overview

This chapter is a blueprint for designing scalable data platforms, in addition to outlining principles, architectures, and technologies that will enable dealing with growing data needs. It starts by defining the four fundamental Scalability characteristics of Applications, namely Vertical and Horizontal Scalability, Elasticity and Consistency, and the importance of these characteristics in today's systems. Discussed is the architecture of patterned state-of-the-art data platforms and how diverse components like data pipelines, microservices, and the visualization layer are systematically integrated in a modular manner to form feasible structures.

This chapter compares SQL databases to NoSQL databases discusses the scalability of each and provides the reader with the necessary information when selecting between a number of databases for different tasks. It also discusses distributed computation platforms such as Apache Hadoop and Spark that can serve large-scale data processing by parallelism. In the case of scalability and hybrid implementation, data warehousing and data lakes are compared.

The giants such as AWS, Azure, and GCP are discussed for their scalability and the cost they possess. Methods for disentangling storage and compute capacity, making the architecture fault-tolerant, and implementing active/ active redundancy and disaster-recovery plans are described. Performance monitoring is the final aspect of platform management discussed in the chapter through stressing the importance of throughput and latency.

2.2 Principles of Scalability in Data Systems

Scalability is among the core concepts for designing data systems so that they progress in terms of storage capacity, utilization, and computational requirements. Elaborated solutions are important for companies because organization management deals with imbalances and increasing loads all the time, especially in terms of big data and the cloud. Here, some considerations on scalability applied to data systems and references to the scientific literature are presented.

1. Horizontal and Vertical scaling.

Horizontal Scaling: This entails a process of placing more nodes or servers so that the workload is shared among two or more facilities. For example, Apache Cassandra, which is considered as a distributed database, directs the data horizontally in the nodes efficiently by using horizontal scaling (Lakshman & Malik, 2010). Horizontal scaling is ideal for systems that need high availability and redundancy as appending further servers would not affect an ongoing process. It also enables systems to more easily absorb increased traffic loads, thus the preferred method for cloud-first applications.

Vertical Scaling: This implies a process of augmenting the facility of a single hardware resource in a system like an increase in CPU, memory, and storage among others. Horizontal scaling is more complex than vertical scaling since it does not need much architectural change but has its drawbacks. The acquisition of high-end interfaces is expensive and there is a limit to how much one can push the capacity of a single host. However, these strange points are compensated with vertical scaling which is still used for applications that have basic functionality or applications that use the paradigm of monolithic applications.

2. Load Balancing

Load balancing is a method used to ensure that no component in a system is overloaded by the amount of information it has to process especially in a period of high demand for information by other components in the system. Other methods are round-robin, least-connection, and IP-hash algorithm used to spread links to the request across the servers. Cloud service providers rely on sophisticated load balancers to handle this kind of traffic, so that response can instantly scale up without any delay (Erl et al., 2020). Load balancing also guarantees for fault tolerance as workloads can be ON healthy nodes and during node failures.

3. Partitioning and Sharding

Sharding or partitioning is necessary in handling huge amounts of data. Subdividing data helps systems enhance retrievals' efficiency and lessen competition because queries become more soluble when data is split into smaller pieces. Sharding techniques of the ranges, hash, and directory kinds make sure that the workload is split equally on the machines (Abadi, 2012). For example, in the database of user information, data can be subdivided by user-ID range so that queries will be distributed evenly among the servers.

4. Elasticity

Elasticity is a close cousin of scalability since it considers the resource usage capacity of a system to up or down size them in line with demand. Elastic systems are especially suitable for the clouds since resources in the clouds may be allocated in a demand basis. This metric is useful in minimizing costs, as it means that organizations only incur costs for the consumables that they employ (Armbrust et al., 2010)

5. Asynchronous Processing

Asynchronous processing increases scalability by uniting activities. Queues can also be used to address requests separately with respect to a particular task meaning that rather than waiting for a particular task to finish, systems can handle more operations concurrently. This principle is used consistently in messaging throughput similar to services such as Apache Kafka and RabbitMQ in the distributed system (Kreps et al., 2011).

6. Statelessness

Statelessness a principle of RESTful architectures means to ensure that client requests are not dependent on other requests. That's because, by removing the need for session persistence, it makes horizontal scaling even easier. Since stateless systems are independent of servers for session management, the stateless systems are easier to replicate and distribute as compared to stateful systems (Fielding, 2000).

7. Monitoring and Auto-Scaling

It is especially important to monitor the systems constantly in order to float bottlenecks and guarantee its performance. Prometheus and Grafana are used to monitor the metrics of systems requiring timely scalabilities to be conducted. Auto-scaling, commonly use in cloud services such as AWS and Azure, depends on the monitoring data to make decision based on pre-determined set of rules or parameters (Jani, 2024).

8. Decoupling of Components

A scalable system divides the entire system into individual units and then makes them loosely bound so that individual components can be scaled as and when required. For example, the uses of the microservices architecture provide scalability through a capacity for each service to be scaled up depending on its usage (Peruccon & Simeone, 2023). It also helps to avoid the prospect of one element dominating the entire system, similar to a bottleneck.

9. Consistency, Availability, and Partition Tolerance (CAP Theorem)

CAP is an acronym that describes the trade-off techniques that exist in a system by balancing between partition tolerance, availability, and consistency. The author, noted that it is difficult the maintain all three in a distributed environment at any one point in time. Slot systems differentiate between availability and partition tolerance; thus, they implement models of eventual consistency. This approach is typical for distributed data storage such as DynamoDB where data updates are synchronized over time (DeCandia et al., 2007).

10. Challenges and Best Practices

It is important to note that scaling is a critical component of every organization but most often many find it difficult to achieve. This is about a system sync mechanism where there are multi-subsystems: the distributed system entails numerous sub-systems that must synchronize in some ways. Delay time in the network, duplication of data, and other mechanisms used in implementing fault tolerance will also pose a challenge to the scalability processes. Some or all of these challenges can be dealt by practicing the modular design, exploiting the ready-made frameworks and prominently testing at a large scale.

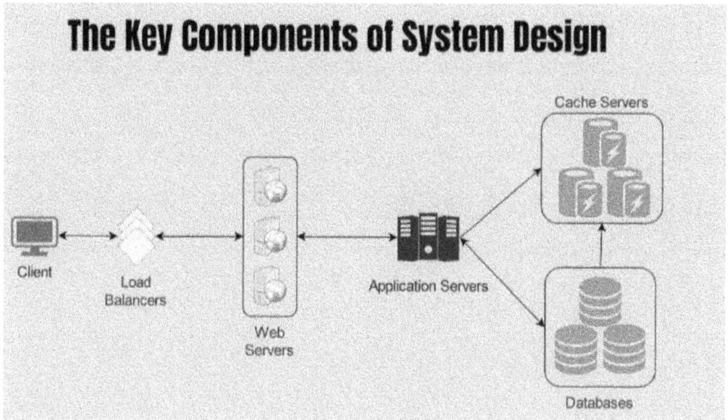

Figure 2.1: System Design Components
Source: - *(Mutiso, 2024)*

The Role of Fault Tolerance in Modern System Design

In addition to scalability, modern systems must be fault-tolerant. Distributed systems will always fail because of software faults, network problems, or hardware failures. A system with fault tolerance can continue to operate properly even if some of its components fail.

1. Redundancy and Failover Mechanisms

- **Redundancy** is the foundation of fault tolerance. By duplicating critical components, systems can survive the failure of individual elements. Failover mechanisms ensure that when one component fails, another redundant component takes over its responsibilities seamlessly.
- **Example**: In a cloud environment, redundant instances of a database can be deployed across different data centres. If one instance becomes unavailable, a failover process switches to another instance, ensuring that the system remains operational.

2. Data Replication

- In distributed systems, data replication is essential for fault tolerance. By distributing data among several nodes or geographical areas, the system can continue functioning even if one node or data centre fails. Replication strategies can be configured to ensure that data consistency is maintained or that the system operates in an eventually consistent manner.
- **Example**: Distributed databases like Amazon DynamoDB and Google Cloud Spanner use replication to ensure that data remains available across multiple geographical regions, even if a failure occurs in one region.

3. Graceful Degradation

- **Graceful degradation** refers to a system's capacity to continue operating in a limited capacity even in the event that some of its components fail, rather than experiencing a total collapse. This approach ensures that the system can still provide essential services while less critical features may become temporarily unavailable.

- **Example**: If a payment gateway service is down in an e-commerce application, the system can still allow users to browse products and add them to their carts, with payment processing deferred until the service is restored.

4. Error Detection and Self-Healing Systems

- Modern systems need to actively monitor for potential failures and respond to them automatically. Error detection mechanisms such as health checks, logs, and monitoring tools can identify failures before they impact users. Self-healing systems attempt to resolve issues automatically by restarting failed processes or shifting traffic to healthy components.
- **Example**: Kubernetes, a container orchestration platform, can automatically detect when a service is unhealthy and restart it or replace it with a new instance. This ensures that the overall system remains resilient despite occasional failures.

5. Eventual Consistency vs. Strong Consistency

- It can be difficult to maintain high consistency in distributed systems, when all nodes have the same data at the same time. This can have a detrimental effect on availability and performance. Eventual consistency offers a more fault-tolerant approach, where the system guarantees that, given enough time, all nodes will have the same data, even if they are temporarily out of sync.
- **Example**: NoSQL databases like Cassandra and MongoDB use eventual consistency to improve fault tolerance and availability in distributed systems (Mutiso, 2024).

2.3   Architecture of Modern Data Platforms

Every engineer's job description, every organization's strategy, and every CIO's hopes and fears all revolve on data. Every day, new data producers and collectors are being constructed. Observability tools and data platforms are the collectors, and users of websites and applications, IoT devices and sensors, and various infrastructure layers are the creators.

This poses a problem. Organisations that use this data find it difficult to absorb, manage, and query its enormous amount, despite the fact that the data itself offers many opportunities and insights. Because they might not have the infrastructure or platforms in place to manage the data at the rate it is created or transmitted, organisations are also facing a growing challenge with data velocity.

Although there are several software solutions available to tackle the difficulties presented by large amounts of data, this presents one of two difficulties on its own. The fact that the current data platform is based on outdated designs or technologies makes it difficult to grow, patch, and adapt to new needs. Organisations are adopting new technology, yet there is a widespread lack of expertise or experience. This presents another difficulty. Although certain technologies are more widely used and dominate the market, organisations typically employ a variety of technologies and processes without a defined plan or direction.

OLTP vs OLAP

Modern data platforms are made to support OLAP systems, which in turn depend on OLTP systems to some extent. Let's look at the distinctions between the two before getting into the specifics and will explain why you need both.

- **OLTP**

The primary function of a contemporary data platform is to facilitate online transaction processing (OLTP). This is often the system's only source of truth, where data entities are generated, changed, and modified. The ACID guarantees of an OLTP system are its most crucial feature. It should offer atomicity, isolation, and consistency guarantees as it is meant to preserve transactions. Obviously, durability is a must because this database is now your only source of information.

Naturally, because they are completely ACID compliant and have been tried and proved over decades of use, classic SQL databases are frequently employed as the OLTP component of the data platform.

OLTP systems are used worldwide to keep track of transaction data in a variety of businesses. They store, record, and retrieve such data in response to queries in a contemporary data platform. OLTP systems may essentially give end users basic information in an easy-to-understand manner. Mutable data, such that kept in OLTP systems, must be dependable by having ACID guarantees since the data on data platforms is always changing.

Data Warehouse and Data Lake Architectures

It's important to utilize the term "data platform." These days, the terms "Data Lake" and "Data Warehouse" are frequently used interchangeably, despite their minor variations in actuality. Usually, these discrepancies relate to the data's maturity and the clarity of the schema that is applied to it. If done correctly, however, they both describe OLAP systems and tend to have comparable architectures. The common characteristics of these solutions will be examined in this section.

- **Data Ingest**

The data that is coming in at the beginning of a data platform is shown on the far left of the above figure. The data will be coming in from a variety of sources, including several OLTP databases, Internet of Things devices and sensors, and many more. It will be ingested in real time, in batches, or in micro-batch.

Streaming ingestion, often known as real-time ingestion, is a popular topic. As previously said, the insights produced are more precise and practical the narrower the gap between your data platform and the OLTP system. Although real-time ingestion data is never really "live," it is more accurate than batch ingestion.

- **Queuing**

The queuing system is the foundation of a data platform. Although Kafka is shown in the figure as the message broker and queuing system, it could be any number of technologies, including AWS Kinesis, Google Pub Sub, Apache Pulsar, and many more.

In between the ETL/ELT engines and the ingested data, the queuing technology serves as a message broker. In order to protect the rest of the pipeline from being overloaded in the case of an unexpected volume or surge in data, the broker buffers data between the two phases. Since micro-batch and real-time ingestion are more vulnerable to unexpected data floods entering the pipeline, this poses a greater danger. When a new version is released or a mistake occurs, it frequently also permits replaying of occurrences.

ETL or ELT Engines

The ETL or ELT engines are an essential component of a data platform once data has been fed into the queue. Although there are readymade software solutions like Up solver, companies with more diverse data and complex big data problems tend to base their pipeline on open-source technologies like Apache NiFi, Airflow, and several more. The goal of both ETL and ELT engines is the same. After merging the data, the final result is stored for the following step in the process. Their methods for producing that final result differ.

Extract Transform Load (ETL) is the most often used paradigm for data processing. The buffered data from many different sources is regimented into structured data by an ETL procedure. The converged data is subsequently written to a destination, usually merely storage that has been optimised for querying.

In contrast to ETL, ELT engines use an interactive technique where the output of each micro-transformation is likewise stored on the target storage after moving all the data onto it. They then perform transformational merging and convergence. Although the cost of storing resources is higher, this makes the entire process auditable.

- **Stream Processing**

In recent years, stream processing technologies have advanced significantly. Alerting, anomaly detection, and other use cases that demand low latency response may now be developed and run directly on the stream of incoming data, achieving excellent performance and accuracy.

Stream processing has made it possible to process traditional use cases that previously needed batch processing. Instead of waiting for the data to go into storage first, technologies like Flink and Kafka Streams calculate data in real-time immediately on the data coming from the queue.

Generally speaking, due to the computing resources needed, it will be significantly more cost-effective to conduct calculations and queries on optimised data stored in storage rather than on the stream (Syn-Hershko, 2022).

The diagram below illustrates the different layers of modern data platforms. Details of each layer are mentioned below,

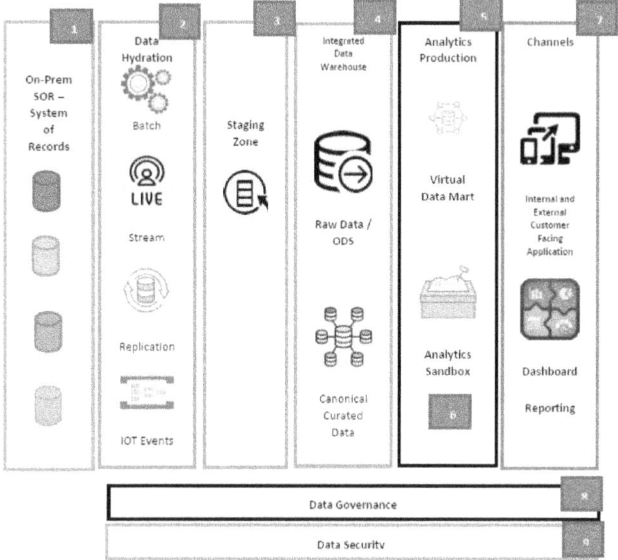

Figure 2.2: Modern Data Platform

Source: - *(Kosambia, 2021)*

- **On-Premise Data:** consists of data in various application databases called system of records (SOR) which today exist on-premise data canter infrastructure. Data will be ingested into the cloud staging zone from on-premise SOR for respective application / Data Sources System of Records.

- **Data hydration:** This layer will move on-premise data to cloud staging zone. From On-premise to cloud staging data hydration we will have 4 ingestion patterns i.e. Batch, Stream, Replication and IOT event consumption depending on use case we have.

- **Staging Zone**: The first landing area of data on the cloud will be in the staging zone. Data will be made available here for faster retrieval. This can also be called as raw / bronze zone with unrefined data in a native format like CSV, JSON, Text binary, XML etc

- **Integrated Data Warehouse**: Data from the staging zone will be ingested into the data warehouse for curation. The ingestion process can use one of the data hydration patterns mentioned above. A curated version of data will be made available to the consumer application in this layer. All analytical workloads will use the curated copy for business ad-hoc analytics. This layer can also be called the Silver zone. Silver will have refined data where data quality standards are applied.

- **Virtual Data Marts**: Application/downstream analytics will use curated copy to create a de-normalized materialized view as per need. This can also be called as Gold Zone. It will have aggregated data sets for downstream systems.

- **Analytical Sandbox:** Will have production quality data for Data scientist teams for exploration and creating, training, and testing ML models. This area can also be called an ephemeral area for exploration and advanced analytical endeavours.

- **Channels**: Analytics and insights will be consumed by various channels for Internal and External facing applications, Dashboards, and reporting tools used by end users

- **Data Governance**: All data since its inception to the cloud will be channelized through a data governance process to ensure data standardization, data classification, data ownership, adding business and technical metadata, and data quality
- **Data Security**: Data will be protected using data entitlement both at a user level and row level, data encryption, tokenization, and masking (Kosambia, 2021).

2.4 Choosing the Right Database: SQL vs. NoSQL

Businesses that offer data-intensive apps have a lot of choices to make about their optimal implementation and upkeep. Choosing the right platforms to store and transport the application data is one of the most important considerations. Due to their capacity to safeguard data and guarantee its integrity, SQL databases were previously chosen by the majority of organisations. However, because NoSQL databases are better able to manage the volume of unstructured and semi-structured data that comes with the internet and cloud technologies, many organisations have resorted to using them.

Even with this tendency, many IT teams still handle more conventional workloads, frequently in tandem with their contemporary apps, and it's not always obvious whether they should use NoSQL or SQL database systems. Both have pros and cons, but they are not the same in terms of construction, data storage, and application access. Additionally, an organisation can only choose the kind that will best fit their workloads in the near and far future by being aware of these variances.

Introducing SQL databases

An SQL database are built on the relational paradigm first presented by Edgar F. Codd in the early 1970s, they are sometimes referred to as relational databases. The relational model lays out a process for classifying structured data into tables with rows and columns, or relations, and establishing the connections between those tables. The relational database quickly became the de facto standard for storing and managing data in organisations of all sizes after its introduction, and it is still a widely used technology today. Of course, relational theory encompasses much more than this, and I have been much looser with its terminology.

The Structured Query Language (SQL), a standards-based computer language used to specify database design and table connections, is the foundation of the relational paradigm. Data from such tables may also be stored, modified, and retrieved using the language. The International Organisation for Standardisation (ISO) and the American National Standards Institute (ANSI) have both approved SQL, which is well-known and extensively used by developers worldwide.

Relational databases have long been used by organisations because they provide a number of significant features that make them well-suited to corporate workloads. Their intrinsic properties, such normalization, atomicity, and consistency, guarantee the integrity of the data across its lifetime, and they are best suited for managing highly structured data. In addition to improving storage efficiency, these capabilities offer standards-based SQL query capability that is versatile.

Relational databases can have certain drawbacks, though. They perform well with organized data but poorly with unstructured or semi-structured data, particularly when used in large quantities. Even for structured data, SQL databases may be challenging to expand horizontally, which makes using them for dispersed big data applications challenging. Additionally, a relational database has a strict schema that needs to be well thought out and is difficult to modify to meet changing needs. This leaves little space for the dynamic nature of many modern applications and development processes.

Despite these difficulties, many businesses continue to choose SQL databases because manufacturers provide a wide range of advanced relational database technologies, including Microsoft SQL Server, Oracle Database, IBM DB2, MySQL, PostgreSQL, and many more.

Introducing NoSQL databases

The proliferation of unstructured and semi-structured data has led many organisations to use NoSQL databases, which are commonly understood to mean "not SQL" or "not only SQL." Even if the name is a little ambiguous, it indicates a class of databases that are more adaptable and scalable than conventional SQL databases. NoSQL databases are not limited to a particular data model like relational databases are, nor do they follow the rigorous schema structure that conventional databases have.

In actuality, there is no requirement that a NoSQL database adhere to any one paradigm, and the market is still sufficiently flexible to accommodate any scenario that could arise. Nevertheless, four fundamental NoSQL models have developed, and each has products available:

- **Key-value databases.** Almost any kind of data may be saved in any format as it is organized using a key-value structure that links distinct identifiers to particular data blobs. Berkeley DB, Amazon DynamoDB, and Redis are a few examples of products.
- **Document databases.** Although not restricted to preset fields or elements, data is saved as documents in formats like JSON or XML, each document having its own unique key. BaseX, CouchDB, and MongoDB are a few examples of products.
- **Column-oriented databases.** Large amounts of data may be swiftly queried and aggregated since the data is stored as strongly-typed columns rather than rows. Other names for these kinds of databases are column-store, wide-column store, and column-family. Cloudera, Bigtable, and HBase are a few examples of products.
- **Graph databases.** In essence, the links between the data are treated as significant as the data itself, and the data is kept in graph structures that specify how the data is related. Examples of products are Infinite Graph and Neo4J.

Depending on the workloads an organisation is attempting to handle, each form of NoSQL database has pros and cons. Furthermore, not every NoSQL product cleanly falls into one of these groups. Azure Cosmos DB, for instance, offers five APIs that increase the service's use. These consist of the Table API, Cassandra API, Gremlin API, MongoDB API, and SQL API.

But regardless of the product, any reliable NoSQL database should have much greater flexibility than SQL databases and be able to manage distributed large data workloads and grow horizontally. Because NoSQL databases are not restricted by strict schema patterns, they can facilitate the beginning of projects by development teams. Nevertheless, NoSQL databases are often unable to provide the same levels of data integrity as relational database systems due to their lack of maturity.

Comparing SQL and NoSQL databases

To better comprehend the differences between SQL and NoSQL databases, it might be helpful to view a side-by-side comparison of the two kinds. The table that follows deconstructs many of the primary differences between SQL and NoSQL.

	SQL databases	NoSQL databases
Data structure	SQL databases are ideal for highly organized data because of the relational paradigm that underpins the SQL data structure, which normalizes data across precisely specified tables and standardizes the relationships between those tables	The NoSQL data structure is adaptable enough to support many models, such as key-value, document, column-oriented, and graph, and does not necessitate a normalized setup or conform to a relational model.
Language	The SQL language is the foundation of SQL databases. While many relational database systems use modified versions of the language, such SQL Server's Transact-SQL (T-SQL), to provide product-specific functionality, some relational database products only support pure SQL. Nonetheless, the fundamental ANSI/ISO language components are supported by all SQL databases.	NoSQL databases are not language-specific. The kind of NoSQL database, the particular implementation, and the particular operation all influence the language utilised. MongoDB, for instance, uses the JavaScript programming language to store all documents in a JSON format..

Schemas	In order to optimized storage and guarantee data integrity, a SQL database needs a predetermined schema that dictates how tables are set up and data is stored. This inflexible structure restricts flexibility.	NoSQL databases provide a great degree of flexibility because they employ a dynamic schema that eliminates the need for a preset data structure. For example, they allow you to add documents with various fields to the same database.
Data integrity	High levels of data integrity are provided by SQL databases, which follow the atomicity, consistency, isolation, and durability (ACID) principles. These characteristics are crucial for workloads like financial transactions.	Most NoSQL databases follow the BASE principles (basic availability, soft state, and eventual consistency), which means that data in a distributed environment may be momentarily inconsistent. This makes it challenging for NoSQL databases to provide the same level of data integrity as SQL databases.
Scalability	SQL databases are not very effective at scaling horizontally, which makes them unsuitable for big, dispersed data sets. Instead, they grow largely vertically, making it simple to scale them up by adding resources like CPUs or RAM.	Large amounts of dispersed data may be accommodated while handling higher traffic volumes because to NoSQL databases' very effective horizontal scalability across systems and locales.

Although comparing SQL and NoSQL databases in this manner might be convenient, the distinctions between them are not always clear-cut. In order to make their goods more ubiquitous, vendors have been gradually adding functionality. For instance, MySQL now has a native JSON data type for storing and verifying JSON documents, while MongoDB now supports multi-document ACID transactions.

How to decide between databases that are NoSQL and SQL

The workloads you want to handle, as well as the type and volume of data, will play a major role in your selection between SQL and NoSQL. Nevertheless, you should also take into account the variations in the database systems themselves, including their stability, maturity, license costs, vendor support, and the size and involvement of the development communities.

Meanwhile, the following table offers some broad recommendations that you may want to take into account while comparing the two types.

Consider SQL databases when...	Consider NoSQL databases when...
• The highly organized nature of your data means that it doesn't change very often. • Accounting and financial applications are examples of transaction-oriented systems that you support. • A high level of data security and integrity is necessary	• The relational model isn't applicable to the vast volumes of unstructured or semi-structured data you're working with. • You need a dynamic schema's flexibility or prefer greater control over the data model. • You need a database system that can grow horizontally, maybe spanning several different regions.

These are only recommendations, once more. You should think about each scenario separately, keeping in mind the workload needs and the form of your data. However, bear in mind that you are not restricted to a single database type. The best of both worlds may be had since many organisations have adopted both SQL and NoSQL database systems to satisfy their various needs. Furthermore, bear in mind that database technology are always changing and that new factors are frequently introduced to the mix.

One thing is certain, though, regardless of how the industry changes: the better you understand your data and the database options available, the more informed you can make when selecting a database system, and the more capable you will be of supporting your workloads in the future(Sheldon, 2021).

2.5 Distributed Computing Frameworks

The core element of distributed computing systems is distributed computing frameworks. They offer a crucial means of facilitating the effective processing of large data on cloud or cluster systems. The rate at which big data is growing outpaces the rate at which clusters' big data processing capability is growing. Therefore, large data analysis activities, which frequently call for executing sophisticated analytical algorithms on terabytes of data, cannot be supported by distributed computing frameworks based on the MapReduce computing paradigm. These frameworks encounter three difficulties when executing such tasks: limited analytical algorithms because many serial algorithms cannot be implemented in the MapReduce programming model; computational inefficiency due to high I/O and communication costs; and non-scalability to big data due to memory limitations. To overcome these obstacles, new distributed computing frameworks must be created. In this study, we examine distributed computing frameworks of the MapReduce type that are already in use for managing large data and talk about the issues they have while analysing large data. Furthermore, we provide a distributed computing architecture that is not MapReduce and might potentially address the difficulties associated with large data analysis.

Basic steps of distributed computing

The divide-and-conquer approach divides a large data set to be studied into smaller data block files, which are then spread among cluster nodes and handled in a distributed file system, such as HDFS. To compute the huge data set on a cluster using the shared-nothing architecture, the two fundamental distributed computing processes are as follows:

1. **Local operation.** To generate the local results, local operations calculate the local data files that are read into the memory of local nodes on each node. Local actions are carried out concurrently throughout the cluster's nodes and independently on each node. In MapReduce, this stage is referred to as the Map operation.
2. **Global operation.** The local results are sent to certain nodes, such as the master node, where the global operations calculate the global results based on the local results. In MapReduce, this stage is referred to as the Reduce operation.

The two fundamental processes of distributed computing are shown in Figure 2. The discs of N nodes house the data block files. Every node is given a local operation to perform on the local data blocks. Each node takes the local data block file into memory and performs operations on it to provide a local output, such as Hadoop MapReduce, which may then be written back to the local disc. To produce the global result, the global operation transmits each node's local result to the master node. There are three components to the overall calculation cost of the two phases.

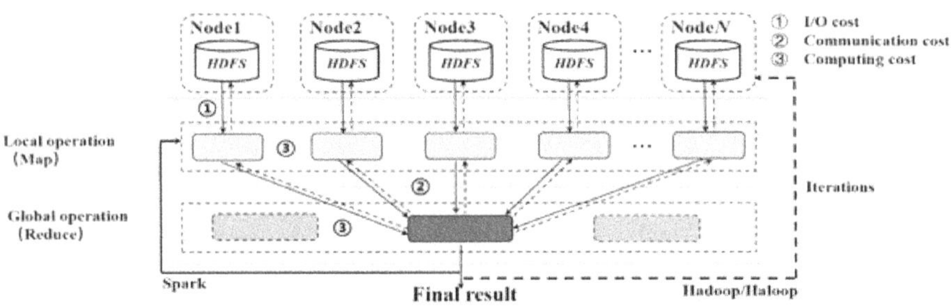

Figure 2.3: MapReduce-type distributed computing framework. Hadoop and Haloop iterative algorithms generate I/O, communication, and computing costs repeatedly, whereas Spark reduces I/O costs significantly by using in-memory computing.

Source: - *(QLICK, 2023)*

1. I/O costs are the expenses incurred by each node when reading and writing local files;
2. Communication costs are the expenses associated with sending local results to the nodes that compute the global outcomes;
3. Computing costs are the expenses incurred when carrying out local and worldwide processes.

In the event when a distributed algorithm completes a computational job, such as counting the word frequencies of web pages, in a single iteration on the dataset, all costs are incurred once. But, if an iterative technique is used to analyze a distributed dataset, the aforementioned expenses will recur in proportion to the number of iterations required to finish the calculation. Therefore, while developing distributed computing frameworks and algorithms for large data analysis, minimizing these costs is a crucial challenge. We go over how Hadoop MapReduce, Haloop, and Spark systems manage the aforementioned expenses in the two fundamental distributed computing processes in the next subsections.

1. Hadoop MapReduce

A semi-standard platform for distributed large data applications is Hadoop. The Hadoop MapReduce computing architecture breaks down an iterative algorithm into a series of mapper and reducer operation pairs that are carried out in a pipeline fashion. A single set of mapper and reducer operations is used as an example.

1. The HDFS-stored data block files are read into local nodes' memory one at a time during the mapper phase. Every local node undergoes the predetermined mapper operation separately and concurrently. The local result is written to the local disc when the procedure is finished. The subsequent reducer operation reads the local result back into the memory.
2. The internal shuffle operation gets the local results from the local nodes prior to the reducer operation. It then shuffles the local data on the reduction keys and temporarily stores the associated

data blocks to the nodes where the reducer actions will be carried out. Then, if the current pair of mapper and reducer operations is not the last, the reducer operation reads the shuffled local result data to the reduce node's memory and works on the data to produce the global results as the intermediate results for the subsequent pair, or as the algorithm's final results if the pair is the last. Communication costs are increased in this stage since the shuffle process necessitates transferring the local results to the appropriate reduction nodes.

It is clear that each pair has high communication costs between the mapper and reducer operations. The shuffle operation will happen often and result in significant communication costs if the mapper and reducer operation sequence has a lot of pairs. Unfortunately, the MapReduce computing paradigm does not allow for the avoidance of this communication cost.

Hadoop MapReduce's I/O cost is especially significant when iterative methods are used since each mapper and reducer operation involves I/O operations. It is not always efficient to build an iterative algorithm in this approach since both mapper and reducer operations are carried out in the {{key, value}} data format. The effectiveness of various algorithms' execution is impacted by this implementation type.

2. Haloop

A Hadoop derivative called Haloop uses the same computation process as Hadoop [91]. It was created with the following changes to increase Hadoop's computation performance on iterative tasks:

- Haloop uses a loop control technique to limit the number of iterations by combining several mapper and reducer operation pairs into a single subtask. This reduces the number of subtasks, with each group being completed in a single iteration.
- To prevent needless network, I/O operations, the invariant data multiplexed in the iterative computation is stored and indexed to local discs.

- The work is planned to use as much local data as feasible in order to minimize the expense of connection.

Haloop improves overall computing speed over Hadoop while only partially lowering I/O, connectivity, and computation expenses with the aforementioned Hadoop improvements. However, because to their partial removal from the computational environment, all three forms of costs remain substantial for complicated iterative tasks over large data sets. Haloop is therefore not frequently utilised in practical applications (X. Sun et al., 2023).

2.6 Data Warehousing and Data Lakes

All of the data, both structured and unstructured, in your company is kept in a data lake. Imagine it as a huge lake-like storage pool for data in its unprocessed, natural nature. Most organisations generate massive amounts of data, which may be handled by a data lake architecture without the requirement for initial data structure. To enable data analytics tools to identify insights that guide important business choices, data stored in a data lake may be leveraged to create data pipelines.

Data Lake Benefits

Competent data scientists or end-to-end self-service BI tools may access a wider variety of data much more quickly than in a data warehouse since the massive amounts of data in a data lake are not organized before being stored.

1. Cost-effective storage is possible for large amounts of both organized and unstructured data, such as call logs and ERP transactions.
2. Keeping data in its raw state allows it to be used much more quickly.
3. A wider variety of data may be examined in novel ways to produce surprising and unobtainable insights.

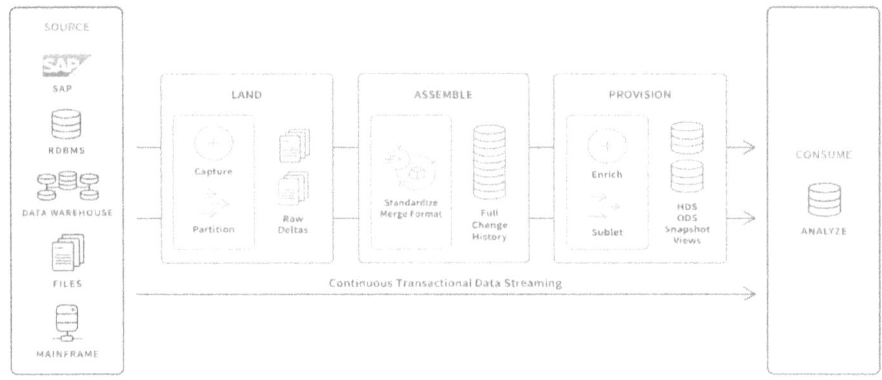

Figure 2.4: ETL data pipelines and schema-on-read transformations

Source: - (QLICK, 2023)

To enable analytics, data science, and machine learning to use data stored in a data lake, data engineers can create ETL data pipelines and schema-on-read transformations. You can get beyond the drawbacks of hand-coded, sluggish scripts and limited engineering resources by using managed data lake building technologies.

These days, a lot of businesses are implementing Delta Lake, an open-source storage layer that improves data lake speed, flexibility, and dependability by utilising ACID compliance from transactional databases. It's especially helpful in situations when your data lake needs transactional capabilities and schema enforcement. It makes it possible to build data lake homes, which facilitate machine learning and data warehousing directly on the data lake. For large-scale datasets, it provides capabilities like data versioning, scalable metadata handling, and schema enforcement, guaranteeing data quality and dependability for analytics and data science jobs.

- **Data warming**

A data warehouse is a storehouse for corporate data, much like a data lake. A data warehouse, in contrast to a data lake, houses only highly organized and unified data to meet certain business intelligence and analytics requirements. Consider it analogous to a real warehouse, where items are first processed before being arranged into sections and placed on shelves (known as data marts). Warehouse data is available for use in supporting reporting and historical analysis to guide decisions across all business divisions of an organisation.

A database that is optimised for scalable BI and analytics and kept as a managed service on a public cloud is called a cloud data warehouse. It allows you to quickly expand or contract your data warehouses to accommodate shifting business demands and budgets, eliminating the limitation of physical data centers.

Data Warehouse Benefits

Organisations may profit much from a data warehouse, particularly when it comes to analytics and business intelligence. Data kept in a warehouse functions as a reliable "single source of truth" following the first cleaning and processing, which is crucial for corporate data analysis, teamwork, and improved insights. A data warehouse has three main benefits, which include:

1. It is much simpler for analysts and business users to access and analyze this data because little to no data preparation is required.
2. Businesses may convert information into insight more rapidly because accurate, comprehensive data is more readily available.
3. A single source of truth is provided by unified, harmonized data, fostering confidence in data insights and cross-business line decision-making.

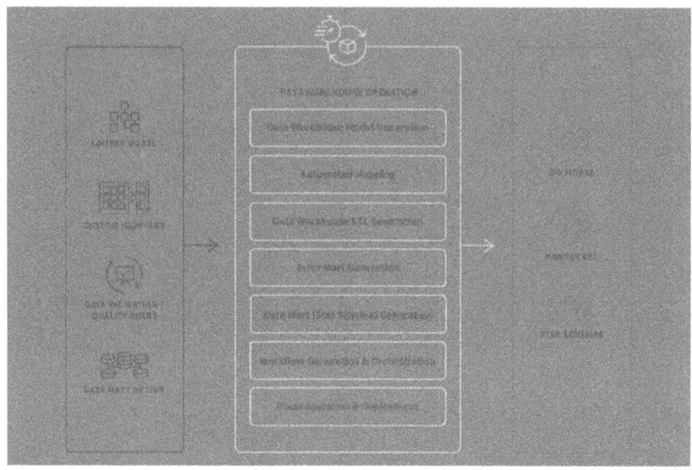

Figure 2.5: Data ware house operation

Source: - (QLICK, 2023)

Data Lake vs Data Warehouse

Data warehouses are the primary tool used by the vast majority of organisations, and cloud data warehouses are clearly becoming more popular. Data scientists usually employ data lakes for flat file exploration and machine learning.

To meet the range of their data storage requirements, many businesses still employ both data lakes and data warehouses. By using a data lake house, some decide to integrate the essential features of each. To give a comprehensive data storage solution for your company, let's compare and contrast data lakes and data warehouses (QLICK, 2023).

2.7 Cloud-Based Data Platforms: AWS, Azure, and GCP

The way businesses manage digital operations has been completely transformed by cloud computing. The three cloud service providers that are leading the global cloud industry are Google Cloud Platform (GCP), Microsoft Azure, and Amazon Web Services (AWS).

The following three major cloud computing companies each provide experience and knowledge to their dependable and feature-rich platforms:

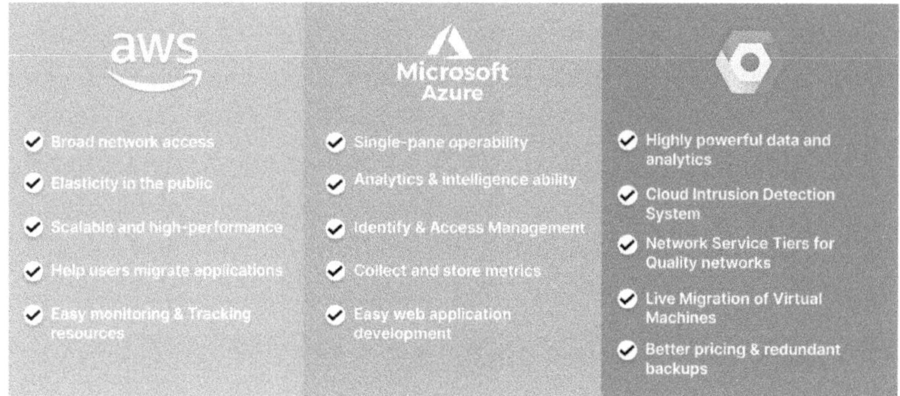

Figure 2.6: AWS vs. Azure vs. GCP: Detailed Comparison

Source: - *(Orangemantra, 2022).*

- **Amazon Web Services (AWS)**

Currently, Amazon Web Services, a division of Amazon.com, Inc., leads the market. This cloud platform is the most established and provides a vast array of services to governments, small and large businesses, and individual developers.

In the beginning, AWS was an internal cloud platform. In 2006, it developed into a publicly accessible on-demand internet computing resource that provided services such as elastic compute cloud (EC2) and Amazon S3 cloud storage. Millions of people may now access more than 200 fully functional services via AWS. According to data released in The Register, it generates one out of every three dollars spent on cloud services, growing at a pace of 37% annually. In 2023, it generated 54% of Amazon's overall operating profits.

- **Microsoft Azure**

Amazon Web Services, a division of Amazon.com, Inc., is now the market leader. It is the most established cloud platform and provides a vast array of services to governments, small and large businesses, and individual developers.

In the beginning, AWS was an internal cloud platform. In 2006, it developed into a publicly accessible on-demand internet computing resource that provided services such as elastic compute cloud (EC2) and Amazon S3 cloud storage. Millions of people may now access more than 200 fully functional services via AWS. According to data released in The Register, it generates one out of every three dollars spent on cloud services, growing at a pace of 37% annually. In 2023, it generated 54% of Amazon's overall operating profits.

- **Google Cloud Platform (GCP)**

GCP is the smallest of the three major cloud providers when compared to AWS and Azure.

Nevertheless, according to The Register, it is expanding at the highest rate, gaining 54% of the market. The business anticipates a 20% growth in 2024.

At the moment, GCP provides more than 200 services covering big data, networking, computing, and other areas. These days, Google Workspace, business Android, and Chrome OS are among the services offered by GCP.

Comparing AWS, GCP, and Azure regions and availability

The first factor to take into account when selecting a cloud provider is the availability and supported geographies. These have a direct effect on your cloud's performance because of things like latency and compliance needs, particularly when handling data.

The Big 3's current position as of September 2021:

- **AWS** has 105 availability zones spread over 33 geographic locations. In the near future, they want to add 21 additional availability zones and seven other regions. They service twelve regional edge caches and more than 600 edge sites.

- **Microsoft Azure** includes 15 areas under development out of 64. They have 126 availability zones in place and are building 37 more. Microsoft has 192 edge sites throughout the world, including four in the US federal cloud.
- **GCP** includes 40 cloud zones, and eight more will be added soon. They have constructed 187 edge sites and 121 zones.

Specific cloud solutions for the government are offered by each of these platforms (government cloud). With data facilities in China, both AWS and Azure provide specialised services that are also aimed at the Chinese market.

It's crucial to keep in mind that each of Azure, AWS, and GCP serves the majority of the world while comparing them. In order to satisfy the always rising need for processing power, all three are also extending their reach by including other areas and zones (Wickramasinghe, 2024).

➢ *Computing*

It comes down to computing alternatives in the AWS vs. Azure vs. GCP debate. Customisation is the main feature that AWS offers in EC2 instances. because among other things, you may join instances, modify regions, and alter storage. But what if you know about computing or are afraid of expensive costs? With Amazon Lightsail, you can get started right now! It offers serverless computing and affordable, reliable virtual private servers. Additionally, the most important thing to keep in mind is that you only pay for the actual time you spend using AWS. The range of instances accessible, including on-demand, spot, reserved, and so on, is another aspect that provides value.

The Azure computing solution, on the other hand, operates on virtual machines (VMs). Using resources such as Resource Manager and Cloud Services to facilitate the deployment of cloud applications.

In the case of Google, virtual machines (VMs) are housed in Google's data centres via its scalable virtualization platform. The best thing is that they are quick to get going, offer a lot of storage capacity, and can be adjusted to meet the demands of customers.

➢ *Pricing*

The first thing that springs to mind when discussing cloud services is cost. Doesn't it? Additionally, caution is required when discussing certain premium cloud service providers. Comparing the three, however, may be difficult at times. Price reductions, promotions, and other pricing models are available.

AWS: To make pricing easier to grasp, AWS offers a reliable cost calculator. However, verify the final pricing again before deciding on anything. because a variety of taxes and fees will be included. For more advice, you can seek help from the AWS cloud computing organisation.

- *Microsoft Azure:* Azure has very high prices. On its dashboard, there is a billing component as well. Additionally, their cloud-based service price calculator is available.
- *Google Cloud:* Millions of users have already come to love Google. Its cost is the most attractive and easy to use. In an effort to attract clients, it tries to undercut the costs that the majority of cloud service providers charge.

In order to draw clients, Google offers additional incentives. To find out what services they provide, use their calculator. Cloud storage, app engines, compute engines, and so forth.

If you're on a tight budget, you might want to consider Google Cloud. However, the finest platform for quality is AWS(Orangemantra, 2022).

2.8 Scaling Storage and Compute Independently

In today's data architectures, the decoupling of storage and compute capabilities has become a guiding architectural concept. Historically, low and high compute were always bounded, therefore when one had to be scaled, the other had to be as well. However, with the introduction of cloud native solutions, organizations can manage these resources with scale, optimize, and balance capabilities, cost, and flexibility. This article analyses the possibility and advantages of decoupled storage and compute in data systems and their pros and cons as well as the strategies of their implementation.

Concept of Independent Scaling.

Loosely coupled storage and computation means that one of these two resources can be adjusted according to its demand independently of the other resource that is available in the given system architecture. This is particularly of significance in the scenarios where storage and computes requirements vary when it comes to growth. For instance:

Storage-Intensive Workloads: Data storage such as data lakes, video archives and IoT's do not necessarily necessitate intense computing but they do demand excessive conventional storage.

Compute-Intensive Workloads: Such workload as machine learning, real time analytics or simulation require a high optimal compute capacity with relatively low storage requirements.

Self-contained sub-services that can be designed independently provide another advantage: technologies like object storage (S3), for example, or serverless computing (AWS Lambda).

Benefits of Independent Scaling

- **Cost Efficiency:** The first benefit of producing nonlinear with independent scales is the optimization of costs. Through loose coupling, organizations are able to prevent themselves from making many of the fatal mistakes which are regularly seen in tightly coupled designs such as the provision of all resources for the organization. For instance, a system that needs more storage can provide the storage as and when needed without having to pay for the extra compute power.

- **Flexibility:** Independent scaling can easily meet different workloads since they have scaling freedom. Organizations are able to flexibly scale computational needs during periods of high activity, for example, just before and during the Black Friday sales period, while not necessarily requiring changes to the storage arrangements.

- **Performance Optimization:** Otherwise known as data architecture, decoupling storage and compute allows systems to excel in certain roles. For instance, layers can be resized to accommodate demand for computational credit during data crunching, when storage configuration can remain constant.

- **Scalability:** Independent scaling supported by cloud platform mechanisms enables organizations to manage big data and use big computations according to their need. This is especially helpful for expanding companies that will need a sizable infrastructure but might not need to periodically completely alter the design.

Key Technologies Enabling Independent Scaling

- **Object Storage:** Technologies like Amazon S3, Azure Blob Storage, and Google Cloud Storage elevate object storage to self-service, scalable, cheap storage that is workload-agnostic in terms of compute. These systems maintain data in a flat namespace which fits well in unstructured data environments.

- **Serverless and Elastic Compute:** Certain serverless computing infrastructure services, like AWS Lambda and Azure Functions, expand and contract in response to load. They enable organizations to perform computational work without strict conformities with physical or logical storage processes.

- **Distributed File Systems:** omputational models like HDFS, Apache Kafka also separate storage and computation, and allow multiple operations at once and lets them scale separately.

- **Containerized Environments:** Container orchestration means, for example, Kubernetes allow for independent scaling of the storage and compute using PVCs and scaling of the compute nodes. This approach is used commonly when the microservice architecture is being adopted.

Challenges of Independent Scaling

- **Complexity:** Marker 3, Decoupling storage and compute based on longevity increases the problem of achieving well-designed architecture. For storage and compute layers to communicate effectively, the developers and the system administrators have the responsibility of designing the systems to achieve this goal.
- **Latency:** Hence, the decoupling of compute nodes and storage systems can potentially raise the latency since network inter dependency must always be accounted for, particularly in a distributed computing paradigm.
- **Data Consistency:** While it is possible for storage and compute to be independent for scale-out, synchronizing data between these layers can be difficult, especially within highly transactional systems.
- **Vendor Lock-In:** This forces customer to be locked in with specific cloud services especially those in the decoupled storage and compute framework.

Best Practices for Scaling Storage and Compute Independently

- **Embrace CNAs:** Boundaries between services logically aligned with the idea of designing systems; by using services which are naturally scalar by default. For example, AWS S3 as a storage model and AWS Lambda as a serverless computing model.
- **Implement Data Tiering:** Organisations should employ data tiering methods in reducing the cost of storage. For example, accessing data often used while other data is stored and accessed less frequently can be classified into high-performance storage tiers and less frequently used archive storage respectively.

- **Monitor Workload Patterns:** Employ workload monitoring tools from AWS CloudWatch or Google Stack driver to have insight of more resources and their usage in order to anticipate oncoming traffic. This ensures that there is efficiency in the storage and compute within the business.
- **API usage for Integration:** Api usage should be employed in such a way that the storage layer and the compute layers can communicate efficiently. For example, to read data at rest for compute purposes, it is easy utilizing AWS SDKs to access S3.

The impacts – both positive and negative – aligned within this priority are specifically pertinent to security and governance.

Use strong security standards that will help reduce data vulnerability in decoupled solutions. Secure data when stored on servers and when transmitted and include rights management to reduce the risk (Bhathal & Singh, 2020).

2.9 Fault Tolerance and High Availability

IT infrastructure solutions that guarantee smooth operations and little downtime must have fault tolerance and high availability. Businesses need a strong IT infrastructure in the ever-changing world of technology in order to support operations and meet the changing expectations of their clients. Selecting the best infrastructure solution for your company requires an understanding of the differences between fault tolerance and high availability.

The capacity of an IT infrastructure to continue operating even in the event of unplanned disruptions or system failures is known as fault tolerance. This can involve network disruptions, software issues, or hardware malfunctions. A system uses redundancy techniques, such as mirroring data storage or putting backup power supply in place, to provide fault tolerance.

On the other hand, by guaranteeing that systems are constantly reachable and functional, high availability reduces service disruptions. Implementing failover systems that automatically shift workloads to alternate resources in the event that primary ones become unavailable helps to manage issues like planned maintenance and upgrades. In order to properly spread tasks over numerous servers, high availability frequently depends on load balancing and clustering techniques.

High availability and fault tolerance have separate functions yet both greatly increase overall system dependability. While high-availability infrastructures place a higher priority on keeping services operational even in the face of planned maintenance or possible bottlenecks, fault-tolerant systems place more emphasis on continuing to function even in the event of unplanned breakdowns. An organization's needs and requirements determine which IT infrastructure solution is best.

For instance, because of strict regulatory requirements for data security and service continuity, businesses in highly regulated sectors like banking or healthcare may place a higher priority on fault tolerance than high availability. Conversely, companies that provide web-based services may be more drawn to high-availability solutions since they have to guarantee end customers constant performance levels and quick response times.

Companies aiming to reduce the risks of hardware or software failures and unplanned outages while maximizing overall system resilience may find that concentrating on fault tolerance and high availability is a useful strategy. By using a hybrid infrastructure approach that combines high availability and fault tolerance, businesses can make sure their IT systems are prepared to manage a range of difficulties and sustain steady service quality standards.

Cloud computing is a prominent example of how fault tolerance and high availability complement one other. In order to guarantee that services continue to function even in the face of planned maintenance or unforeseen interruptions, cloud providers usually employ extremely sophisticated infrastructure designs that include redundant components, load-balancing techniques, and automated failover capabilities. This gives businesses a dependable, adaptable IT infrastructure solution that meets their unique requirements for high availability and fault tolerance.

Let's take a closer look at fault tolerance and high availability.

<u>Understanding Fault Tolerance</u>

Anyone working with a computer system has to understand fault tolerance, especially those in charge of overseeing and preserving the integrity of data transmission and storage.

The capacity of a system to continue operating properly even in the event that one or more of its components malfunction or encounter faults is the fundamental definition of fault tolerance. Today, when system dependability and data protection are critical, this idea is essential.

Comparing fault tolerance to redundancy is one approach to examine its meaning. Duplicating essential parts or functions inside a system so that another may take over in the event of a component failure is known as redundancy. A common component of a fault-tolerant design approach is redundancy. It is one way to assist achieve fault tolerance, but it is not the same as fault tolerance.

Even though the phrases may appear identical at first, knowing the differences between fault tolerance and redundancy enables us to recognized their main uses. Additional copies of physical parts, such disc drives or power supply, may be included in a redundant system to guarantee that a backup will be accessible in the event that any one component fails.

Fault-tolerant systems, on the other hand, are built with resilience in mind from the beginning; they use a variety of techniques beyond basic redundancy (such error-checking algorithms), which combine to maintain system functionality even in the face of errors or failures.

There are fault-tolerant systems in many different sectors that need high levels of dependability and data security. The servers of financial organisations and flight control systems, for instance, frequently use these systems as accuracy and uptime are essential to their operations.

Developers may use a variety of techniques to create fault-tolerant systems, each having advantages based on particular needs and applications, such as:

- The "n-version programming" approach entails producing several separately designed iterations of a software module. Within the system, these several versions are then operated simultaneously on different hardware. The other versions can continue processing in the event that one version fails or encounters an issue, preserving system operation as a whole.
- Including self-healing mechanisms in systems, which entail keeping an eye out for deterioration in components and replacing them before they fail, or using modular architectures, which make it simple to replace malfunctioning parts without interfering with the operation of the entire system.

Regardless of the methods used, cost, complexity, and performance trade-offs must all be taken into account when building fault-tolerant systems.

The implementation of high degrees of fault tolerance might occasionally necessitate large time and financial commitments, which should be kept in mind when developing systems that can tolerate failures and carry on without interruption. Compared to simpler systems, these systems may also require more frequent maintenance and monitoring. Additionally, when developing the overall architecture, it is important to take into account the potential performance effect of fault tolerance techniques.

Businesses and people may make more informed strategic decisions about system design, implementation, and maintenance by having a solid understanding of fault tolerance and its different techniques. Businesses can improve the security of their data protection initiatives while guaranteeing optimal performance and reliability by incorporating redundancy with other techniques like modular architectures or error-checking algorithms.

This will ultimately result in increased satisfaction for end users who depend on these vital systems on a daily basis (Highleyman et al., 2005).

Understanding High Availability

Another important idea in the field of technology and computers is high availability. It refers to creating and putting into place procedures and systems that guarantee uninterrupted operation with little downtime.

High-availability solutions try to prevent possible system failures before they happen in order to offer continuous access to data, apps, and services. Load balancing, redundancy, fault tolerance, and other strategies that cooperate to sustain peak performance are used to accomplish this.

Cloud computing offers scalable, dependable, and affordable solutions that can adjust to shifting demands, allowing businesses to provide and utilize high-availability systems. Apps, data storage, and other services are guaranteed to be available even in the event of hardware malfunctions or system outages thanks to high availability cloud computing. This is made feasible by several redundancies, which duplicate important parts of a system so that another may smoothly take over in the event that one fails.

Several elements contribute to the development of a high availability cloud environment, including:

- **Virtual servers.** To provide high availability, they are frequently housed on several physical servers. The workload may be quickly moved to another server without affecting the end user experience in the event that one physical server malfunctions or goes offline for maintenance. This is accomplished with virtualization technology, which makes it possible for several virtual servers to operate on a single physical server, resulting in an architecture that is more scalable and efficient.
- **Monitoring.** Early detection of possible problems, including bottlenecks or resource limitations, allows for the implementation

of preventative actions before they become more serious and cause downtime.

- **Load balancing.** In order to keep any resource from getting overloaded or overwhelmed, this entails dividing the workload among several resources (like servers). It also helps to keep the system stable and maximizes the use of available resources. In order to guarantee high availability, load balancers are usually used to divide incoming network traffic across several servers according to different algorithms and settings. This ensures that each server distributes the load equally.
- **Failover mechanisms.** These procedures guarantee that another system component will automatically take over in the event of a failure or problem without affecting end users. Hardware, network, and application layers are only a few of the infrastructure tiers where failover can occur.

Achieving high availability necessitates careful planning and management in addition to these technological factors, including:

- Creating thorough backup plans to handle possible disruptions or calamities that might negatively impact system operations
- Consistent testing and upkeep of systems to guarantee they continue to function at their best and provide the anticipated levels of dependability and performance

To comprehend high availability, one must acknowledge its significance in today's technologically advanced corporate environment. In high availability cloud computing systems, it entails putting in place redundant, fault-tolerant components and using load-balancing strategies and failover methods to keep things running continuously even in the case of unforeseen events or breakdowns. Organisations can guarantee continuous access to critical applications and services by giving planning, monitoring, and prompt action top priority when managing high availability systems. This will eventually help them succeed overall in an environment that is becoming more and more competitive (Highleyman et al., 2005).

2.10 Performance Metrics and Monitoring

Understanding the condition of your infrastructure and services is crucial to their dependability and stability. You may have a deeper understanding of your infrastructure with a top-notch monitoring system.

Monitoring your system's performance enables you to make adjustments to your infrastructure to prevent future issues in addition to resolving existing ones. An efficient monitoring system gathers, combines, saves, and displays metrics while warning you of any issues with your systems.

Metrics are the fundamental values that are used to evaluate different aspects, comprehend past trends, spot faults and issues, and uncover patterns and anomalies.

Metric Fire is a software as a service (SaaS) tool that gathers, saves, and visualizes data while enabling analysis and use. Enrol in the Metric Fire demo if your organisation has a large amount of data, and we will assist you. To discover what the product can achieve, you may also take advantage of the free trial!

This article will examine the definition of metrics and the many kinds of metrics that are available. We'll also go into great detail on what alerts are and what monitoring is in general.

Key Takeaways

1. The reliability and stability of services depend on understanding infrastructure and services through effective monitoring.
2. A good monitoring system involves data collection, storage, aggregation, visualization, and alerting to identify issues and trends in your systems.
3. Metrics are raw data collected from various sources like hardware, applications, or websites, providing information about resource usage, performance, or user behavior.
4. Metrics can be categorized into subclasses like host metrics (CPU, memory, disk), application metrics (response time, error rates),

network performance metrics (packet loss, throughput), server pool metrics, and external dependencies metrics.

5. Metrics should be clear, collected at suitable frequencies, and aggregated appropriately to gain reliable and useful insights.

What is Monitoring?

The process of turning streams of unprocessed data into information that we can utilize is called monitoring. Metric monitoring allows us to assess how well our systems are working, react to specific occurrences, and spot trends and abnormalities. The process of gathering, combining, and evaluating data in order to increase knowledge of the traits and actions of your components is known as monitoring.

The monitoring system has many functions including:

Data collection. Thousands of metrics may be gathered from many sources by an effective monitoring system. To prevent data loss and enable scalability, it must accomplish this well.

Data storing. Metric values that reflect the present are helpful, but it is nearly always more helpful to compare these figures to historical values in order to comprehend the context of changes and trends. This implies that the monitoring system has to be able to manage and retain data for certain time periods so that older data may be sampled or aggregated.

Data aggregation. The process of collecting data and summarising it is known as data aggregation. In order to compile the data from several sources into a summary for data analysis, the data may be collected from several sources. Additionally, raw data may be combined over a predetermined period of time to provide statistics like count, total, minimum, maximum, and average.

Data visualization. By creating tables and examining their data, metrics may be tracked. However, using various graphs and visualizations to identify patterns and comprehend how various elements fit together is more effective. Numerous visualisation features are offered by good monitoring software. By looking at the display, you can comprehend how several variables interact or how the system changes.

Alerting. Users can get notifications from the monitoring system when certain events occur or when monitored metrics reach predetermined thresholds. This enables you to attend a significant occasion even if you are unable to attend work.

Metric Fire is one of the monitoring solutions that offers all of the aforementioned features and more. Our device allows you to get a comprehensive understanding of your surroundings with little setup. Please schedule a demo with us or sign up for the free trial right now if you'd like further information.

Metrics: definition and types

Metrics are unprocessed data that can be gathered from a variety of sources. Hardware, sensors, apps, webpages, and more may all be considered sources. Examples of the data generated by these sources include user behaviour, performance, and resource utilisation. The operating system may supply this information, or it may be higher-level data types connected to a particular feature or component activity, such the quantity of users on a website or the speed at which a page load.

In general, measurements are gathered periodically, for instance, once every second, once every minute, or at any other interval, contingent upon the attributes of the indicators and the objectives of the metrics monitoring.

Many sources produce metrics, and gathering them is simple. You may accomplish this with little extra effort and yet get substantial rewards by constructing a basic monitoring system.

Operating systems produce metrics - With a single click, operating systems may provide you with a wealth of information. Data, such as CPU utilisation, available disc space, or spent RAM, will be simple for you to obtain. Naturally, the difficulty is in how to process all of that information.

Numerous database servers, web servers, and other software applications, including Learning Management Systems (LMS), generate metrics that may also be gathered. You may set up your apps to generate the metrics you require and have your monitoring system gather them(Smartbear, 2022).

2.11 Chapter Conclusion

The development of customizable data platforms is an important process in the modern world, where the capacity of information processing and constantly changing users' requirements is highly valued. This chapter has discussed the main concepts and the design patterns that are fundamental to building scalable, available, and high-performing big data systems today. Knowing the pros and cons of SQL and NoSQL databases as well as implementing popular distributed computing platforms and using the advantages of cloud computing platforms like Google Cloud Platform, Microsoft Azure, and Amazon Web Services, businesses can create powerful platforms to satisfy business requirements.

The ability to separate the storage and compute domains, the goal for fault tolerance, and the reliability of high availability that comes from redundancy and disaster recovery should never be underestimated. They do not only improve the system and its stability but also train a platform for contingencies. Thirdly, performance monitoring and optimization through metrics like good throughput, latency and consumption of resources guarantees durability in as systems grows.

Thus, the application of these aspects and recommendations allows developing data platforms, which not only scale but also adapt and sustain at a reasonable cost. They allow organizations to extract all the value out of it for innovation, decision-making, and for their sustainability in an environment that is rapidly becoming so competitive.

Multiple Choice Questions (MCQs)

1. Which scaling method involves adding more nodes to share the workload?

    a. Vertical scaling
    b. Horizontal scaling
    c. Elastic scaling
    d. Linear scaling

2. Load balancing helps:

    a. Store more data
    b. Optimize resource usage during high demand
    c. Transform data into visual reports
    d. Validate data quality

3. What does sharding in databases achieve?

    a. Splits data into smaller, manageable parts
    b. Encrypts data for security
    c. Merges multiple datasets
    d. Optimizes SQL queries

4. Elasticity in a cloud system allows:

    a. Fixed resource usage
    b. Dynamic allocation of resources
    c. Static data storage
    d. Limited scalability

5. Stateless systems are advantageous because:

    a. They require session persistence
    b. They simplify horizontal scaling
    c. They rely on centralized servers
    d. They limit data accessibility

6. What is a benefit of microservices architecture?

    a.   Allows independent scaling of services

    b.   Reduces system modularity

    c.   Increases dependency among components

    d.   Centralizes data flow

7. Which caching strategy determines how long data is retained?

    a.   Read-write ratio

    b.   Time-to-live (TTL)

    c.   Query optimization

    d.   Data indexing

8. Active/active redundancy ensures:

    a.   Consistent data latency

    b.   Continuous system availability during failures

    c.   Reduced system throughput

    d.   Delayed fault tolerance

9. What is the main focus of the CAP theorem?

    a.   Consistency, availability, partition tolerance

    b.   Data transformation, serving, and ingestion

    c.   Data governance and visualization

    d.   Security, encryption, and performance

10. Distributed systems improve fault tolerance by:

    a.   Relying on single-node databases

    b.   Implementing data replication

    c.   Using relational databases only

    d.   Avoiding data partitioning

Answer

1	2	3	4	5	6	7	8	9	10
b	b	a	b	b	a	b	b	a	b

REAL-TIME ANALYTICS FUNDAMENTALS

3.1 Chapter Overview

This chapter focuses on the phenomena, approaches, and uses of real-time analytics. It starts by differentiating real-time processing from batch processing while emphasizing its capability to process data and offer insights and decisions at a millisecond level, which will allow businesses to take appropriate action. To this end, this chapter explores the architecture of real-time systems and the Lambda, Kappa, and Delta models as the way to achieve speed and scale.

Real-time analytics applications are presented in the context of different industries, including e-commerce, healthcare, and financial services, as well as manufacturing, to demonstrate its effectiveness in various contexts, including decision-making and customer experience, as well as security. The chapter also discusses some of the issues that arise when implementing real-time solutions including; high data velocities, low latency requirements, and integration issues, in addition to discussing the ways to address them. In addition, the analysis of tools such as Apache Kafka, Amazon Kinesis, and Google Cloud Dataflow presents technological trends behind real-time analysis. We take fraud detection, recommendation systems, and supply chain management to illustrate the idea that real-time analytics generates new business value and enhances organization performance.

3.2 Understanding Real-Time Data Processing

The process of turning unprocessed data into a format that can be used in practical applications is known as data processing. Data gathering, refining, and storage are the phases involved. Data may be processed in a number of ways, including batch and real-time. Real-time processing enables quick data manipulations, insight production, and decision-making, whereas batch processing causes delay.

An extensive description of real-time data processing is given in this article. It walks you through its applications, difficulties, architecture, and tools. Get a competitive edge in your industry and increase the operational efficiency of your business by using this article.

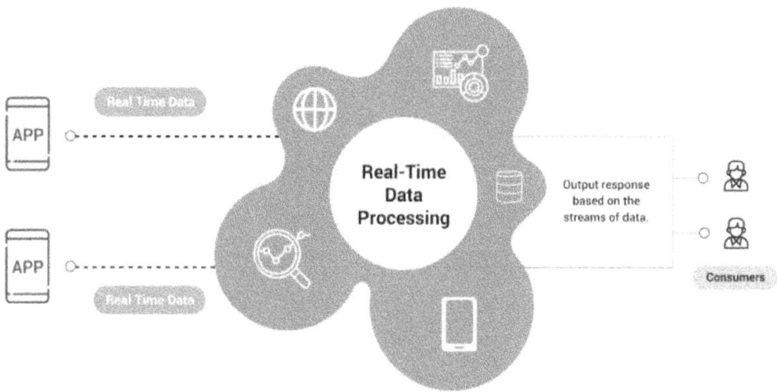

Figure 3.1: Real Time Data Processing

Source: - *(Airbyte, 2024)*

The act of instantly ingesting, transforming, storing, and analyzing data as soon as it is created is known as real-time data processing. With latency in the millisecond range, it is the recommended technique for carrying out speedier data operations. Real-time data processing may be used to speed up data analysis and the creation of business insights.

The algorithm used by e-commerce websites to propose products is a typical example of real-time data processing. Here, you can instantly get information on how customers are using your e-commerce platform, including their search queries, browsing history, clicks, purchases, and add-to-cart activities. After that, you may use this data to provide real-time, more individualized product recommendations to your clients(Silva et al., 2018)inspired by the tremendous growth of connected heterogeneous devices, has pioneered the notion of smart city. Various components, i.e., smart transportation, smart community, smart healthcare, smart grid, etc. which are integrated within smart city architecture aims to enrich the quality of life (QoL.

The following steps make up real-time data processing:

Data Collection: In real-time data processing, rapid data intake is the initial phase. To achieve this, data is gathered from a variety of sources, including social media feeds, server logs, and Internet of Things devices. You may use data streaming solutions like Amazon Kinesis or Apache Kafka to absorb this data.

Data Processing: In this step, the ingested data is combined, cleaned, converted, and enhanced. This transforms information into a format that may be used with other programs and data systems.

Data Storage: Data can be stored in destination systems such relational databases, streaming platforms, or in-memory databases after processing is complete.

Data Distribution: To make the processed data available for use in subsequent processes, it might be dispersed throughout many systems.

The following are some benefits of real-time data processing:

Faster Decision Making: Immediately following analysis, real-time processing gives you data insights. You may use these insights to determine which metrics you are doing well on and which ones are impeding your progress. With this data, you may quickly modify your product's characteristics to better suit the tastes of your target market. This provides you with a competitive advantage and fosters business expansion.

Enhanced Data Quality: Discrepancies in your datasets may be swiftly identified during real-time data migration. Early detection enhances the quality of your data and allows you to eliminate mistakes right away. Additionally, with real-time data flow, mistakes are detected closer to their source. This aids in locating the underlying cause and promptly fixing data errors.

Elevated Customer Experience: Real-time data processing in an organisation may greatly enhance the quality of customer support. You can quickly analyze your consumer data and find the gaps that are deterring customers from purchasing your goods or services. These assessments assist you in refining your marketing plans and product development to boost sales and consumer engagement.

Increased Data Security: Real-time data processing and monitoring can help you identify fraud or security breaches fast. This is very helpful in the stock markets or financial industry. Real-time processing may also be used to spot early warning indications of unfavorable patterns that can affect stock or market values. This enables you to minimize any losses by taking preventive measures in advance.

Immediate Insights and Decision Making: Business executives may react to events and changes as they occur because real-time processing allows for prompt insights and choices based on the most recent data. For instance, real-time processing can assist in identifying fraudulent financial transactions and immediately halting losses in the banking industry, where making judgements fast is crucial.

Real-time processing in retail can be used to monitor supply chain and logistical operations in real time, provide consumers with personalized recommendations based on their browsing and purchase histories, or maximize marketing efforts by focussing on the right people at the right time.

Additionally, real-time processing can facilitate team collaboration and instantaneous communication. This enables businesses to react to shifting consumer demands and market situations faster.

Improved Data Quality and Accuracy

Additionally, real-time processing increases the accuracy and quality of data, enabling businesses to identify and address mistakes and anomalies instantly.

This is particularly important in sectors like healthcare, where even little mistakes and processing hold-ups can have significant consequences. Real-time processing may monitor prescriptions, identify and correct errors in patient data, and assess how well medical equipment are working.

In manufacturing and retail, real-time processing can also improve data accuracy. Real-time processing allows you to:

- Identify and correct errors in client records
- Observe how many things are in stock.
- Monitor the performance of the production equipment.

This improves the quality and completeness of the data, allowing for improved decision-making, increasing customer satisfaction, and lowering mistakes and losses. Using real-time data processing and consumption can help businesses make better decisions and reduce risk.

Enhanced Customer Experience

Customer support systems can also leverage real-time processing to enhance the user experience. Real-time processing enables businesses to assess customer data instantly and produce pertinent, interesting suggestions.

Real-time processing in retail enables companies to track client preferences, offer tailored products and services, and personalise marketing, products, and services to individual customers. By providing personalized suggestions and experiences, businesses may increase sales and income.

Additionally, real-time processing may be used to monitor a client's usage of a website or mobile application, identify and address issues in real time, and provide tailored assistance.

Real-Time Monitoring and Control

Real-time system control and monitoring are made possible by real-time processing. Instantaneous processing and analysis of data enables organisations and agencies to identify issues and respond quickly.

In manufacturing, for instance, it lets processes be watched and managed in real time. Manufacturers can find and fix problems with their tools or processes quickly by collecting and analyzing data. This cuts down on downtime and boosts output and efficiency.

Monitoring and controlling the systems that make and send energy can also be done in real time. So, energy companies can find and fix problems as they happen. This can make energy systems more reliable and stable, which is good for both businesses and customers.

Enhanced Security and Fraud Detection

Businesses may detect fraud much more easily and strengthen the security of their system with real-time processing. Businesses may avoid data breaches and other security issues by processing and evaluating data in real-time.

Additionally, it can assist in real-time network traffic monitoring to identify and stop security threats. Businesses may identify odd patterns of behaviour that can point to a security issue and stop a breach by using real-time processing.

Figure 3.2: Benefits of Real time Data Processing

Source: - *(Richman, 2024b)*

The following categories apply to the real-time processing architecture:

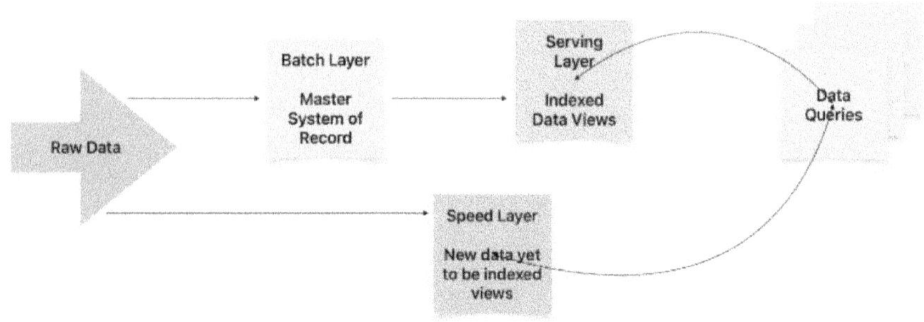

Figure 3.3: Lambda

Source: - *(Airbyte, 2024)*

There are three levels in the Lambda architecture: batch, speed, and serving layers. Data may be processed in batches after being stored in raw form thanks to the batch layer. Hadoop Distributed File Systems (HDFS) and other distributed file systems may be used to store data. Tools like Apache Spark or Apache Flink may be used to process your data in batches.

The speed layer makes it possible to use stream processing technologies like Apache Kafka or Apache Storm to process real-time data in a distributed manner. Batch and speed layer outputs can be combined using the serving layer. It serves as a go-between for the processed data and the end consumer. The serving layer stores processed data in a database like Apache Cassandra or MongoDB and allows you to query the data using query engines like Apache Hive.

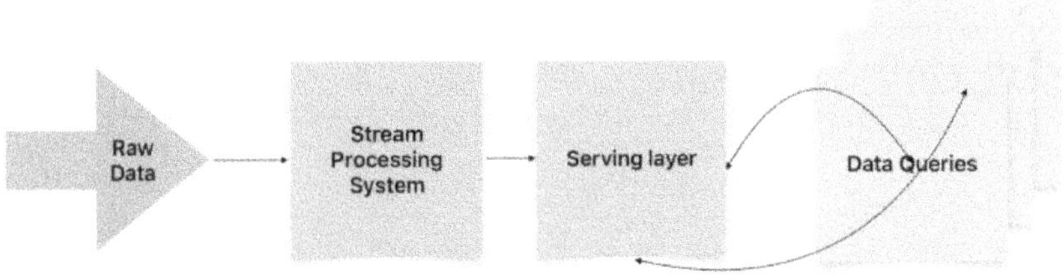

Figure 3.4: Kappa

Source: - *(Airbyte, 2024)*

Compared to Lambda architecture, Kappa architecture is more straightforward and more efficient at handling real-time processing. There is just one streaming layer in it. The Kappa architectural layer ingests, processes, and stores data in a database like Apache Cassandra using technologies like Apache Flink or Apache Kafka stream.

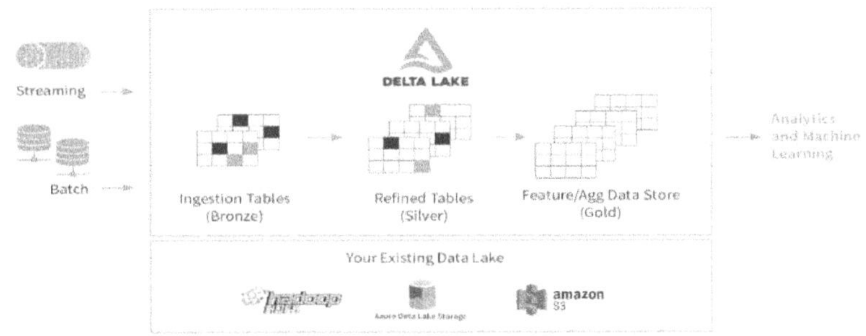

Figure 3.5: Delta

Source: - *(Airbyte, 2024)*

The micro-batching approach in the Delta architecture allows you to combine and optimize the processing and storage capabilities of both Lambda and Kappa. Many contemporary data lakes, such as Delta Lake, use this intermediate technique as the foundation for data processing.

Real Life Applications of Real Time Processing:

a. Financial Trading

The micro-batching approach in the Delta architecture allows you to combine and optimize the processing and storage capabilities of both Lambda and Kappa. Many contemporary data lakes, such as Delta Lake, use this intermediate technique as the foundation for data processing.

Real-time market data may be analyzed using machine learning classifiers that use high-quality real-time data to forecast stock values. A few seconds of delay can have a significant impact on a trade's result in markets that move quickly. Real-time processing gives traders a competitive edge by enabling them to make better informed, strategic decisions.

b. Fraud Detection

Real-time processing enables businesses to spot and respond to questionable activity right away. Real-time data analysis enables businesses to set up automated warnings for odd activity or unexpected expense rises.

An event hub, for instance, can be used in the banking sector to keep an eye on current financial transactions for indications of fraud. The event hub's real-time data analysis enables it to notify security staff of any suspicious behaviour so they can promptly look into it and stop any financial loss.

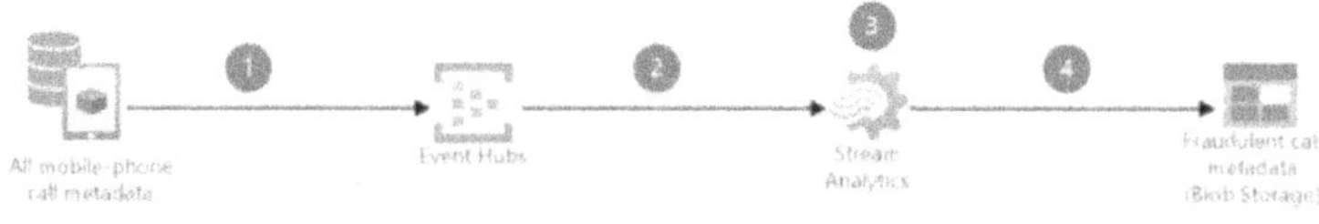

Figure 3.6: Fraud Detection by Real Time Data Processing

Source: - *(Richman, 2024b)*

c. Healthcare

Data from medical systems and electronic health records may be managed in the healthcare industry through real-time processing. Healthcare professionals may swiftly get patient information and base treatment decisions on it by processing and evaluating real-time data.

The data from a patient's vital signs monitor might be continually analyzed using a single-process algorithm, which would notify medical personnel if any readings deviate from the norm. This enables prompt action and may possibly save a patient's life or avoid major consequences.

d. Transportation

Vehicle tracking and route optimisation are made possible by real-time traffic control systems that process transportation data. Real-time traffic reports and suggestions for other routes may be provided to drivers using the vast amounts of GPS data that tracking devices and mobile phones provide.

Radar systems and public transportation routes and timetables can benefit from real-time processing architecture, which also enhances the user experience. Additionally, it may be used to inform travelers of arrival timings and delays, which will improve their trip planning.

e. Security and Surveillance

Real-time processing may significantly enhance an organization's security and surveillance systems by enhancing its capacity to identify threats. Security camera video feeds may be analyzed and suspicious behaviour reported using real-time data and deep learning algorithms.

Real-time processing is used by border and airport security to monitor individuals and potential threats. These systems ensure the safety of people and facilities by promptly responding to potential threats through the analysis of real-time data.

f. ECommerce

Real-time processing is a game-changer for online retailers. It enables companies to take advantage of state-of-the-art e-commerce platforms that provide tailored product suggestions based on browsing activity, precisely control inventory levels, and identify potentially fraudulent transactions as soon as they happen.

3.3 Difference Between Batch and Real-Time Processing

To handle data-driven processes, two of the most effective data processing methods are batch and real-time. The optimal option for an organization's workload depends on its overall strategy and business needs.

Data may now be immediately analyzed from terabytes to petabytes thanks to ongoing technical advancements. The fast proliferation of data has made organisational data management a top priority. IT executives are relied upon by organisations to consistently provide solutions that expedite the accurate and smooth processing of vast amounts of data. The question of whether data processing method is optimal for a given use case is becoming more and more complex as technology and consumer needs change over time.

One paradigm that makes it possible to process large amounts of data at once is batch processing. It is a highly effective method of carrying out extensive operations in parallel, such as sorting, processing, and counting the collected data. Contrarily, real-time data processing entails continuous input, processing, and output of data that has been obtained from the source in a matter of milliseconds. Instantaneous processing of the input data yields an automatic response based on data streams. Since real-time data processing receives data in milliseconds, the master file or database is continuously updated.

Although both batch and real-time procedures are essential, they vary in a few key ways. When all of the data is received, it is saved and processed as a batch over a certain amount of time. A single transactional file is created from this processed batch and kept there until all of the source data has been retrieved. For debugging efficiency, batch processes make sure that big jobs are finished in manageable chunks. Since real-time processes are implemented on systems that must react swiftly and smoothly, they are carried out on the fly as opposed to batch processing. The primary distinction between the batch and real-time processing paradigms is that whereas batch-based processes can be delayed or stopped as needed, real-time processes must react immediately.

Real-time data has an advantage over batch processing as it has become a popular trend as technologies like analytics grow more and more reliant on real-time replies. While real-time data processing takes milliseconds, batch processing has a substantially greater latency, with outputs taking anything from minutes to days, depending on the data batch cycle. Another challenge with batch processing is storage, as big batches need a lot of storage capacity to collect data within a certain time frame. On the other hand, because real-time data processing processes random data supplied instantly at a random moment, it uses less storage (Shahrivari, 2014)

Table 3.1: Difference between Batch Processing System and
Real Time Processing System

Batch Processing System	Real-Time Processing System
The processor in batch processing only has to be active when tasks are allocated to it.	Real-time processing requires processors to be always active and extremely responsive.
Tasks with comparable needs are grouped together and processed by the computer collectively.	This system accepts and processes events that are mostly outside of the computer system by specific dates.
Batch processing does not depend on completion time.	In real time, the task's completion time is crucial.

It offers the most cost-effective and straightforward processing technique for commercial applications.	To manage intricate operating system applications, specialised hardware and software are needed for sophisticated and expensive processing.
Batch processing can also be used with standard computer specifications.	Processing requirements in real time high-end hardware specifications and computer architecture.
There is no time restriction for this procedure.	It must complete a procedure within the allotted time; else, the system would malfunction.
It is focused on measurements.	It is focused on events or actions.
Sorting is done in this system prior to processing.	Sorting is not necessary.
This system collects data for a predetermined amount of time and processes it in batches.	Allows for random data entry at various times.
Batch processing is exemplified by credit card transactions, bill generating, operating system input and output processing, etc.	Real-time processing includes things like customer service, radar systems, weather predictions, temperature readings, and bank ATM transactions.
Large data sets should be processed in chunks, either overnight or according to a timetable.	Handle data in real-time or almost real-time as it comes in.
Increased latency because data is handled in groups following a lag.	Reduced latency because information is processed instantly or with little delay.
Reduced price per data unit since processing is done in batches.	Higher cost per data unit because real-time or almost real-time processing is required.
Perfect for jobs like creating reports, doing nightly data backups, and conducting extensive data analysis.	Perfect for jobs like real-time monitoring, sensor data processing, and fraud detection.

Source: - *(GeeksforGeeks, 2024a)*

In conclusion, data is gathered and processed at a predetermined time using batch processing. Throughout this procedure, the user is unable to communicate with the system and must respond immediately. Real-time processing is employed when prompt processing is necessary to obtain answers. As a result, real-time processing is employed when obtaining immediate processing is necessary for applications such as real-time analytics, whereas batch processing is helpful when immediate action is not needed.

3.4 Tools for Real-Time Analytics

Software programs that allow businesses to gather, process, and evaluate data in real-time or almost real-time are known as real-time analytics tools. These tools give businesses fresh insights into their operations, customer behaviour, and overall business performance, helping them make more informed decisions.

In today's rapidly evolving business environment, traditional data processing methods often lag behind. Real-time analytics tools address this challenge by processing data as generated, allowing for swift, data-driven decisions. They utilize various technologies, including in-memory computing and real-time data rendering, to analyze data streams continuously.

There are several reasons why Real-time streaming is essential in today's data-driven world:

1. Real-time Insights

Real-time streaming empowers businesses with immediate access to their data, allowing for rapid and informed decision-making. This talent is essential in a fast-paced setting where success can be greatly impacted by quick decisions. With real-time insights, organizations can adapt strategies on the fly, optimize operations, and enhance customer interactions.

- **Example:** E-commerce businesses can use real-time data to personalize product recommendations, ensuring that suggestions align with the latest customer behaviour and preferences.

2. Faster Response Times

Real-time streaming accelerates responding to events and incidents, vital for mitigating risks and maintaining high performance. Real-time data processing and analysis enables companies to promptly resolve problems and make the required corrections. This proactive approach helps minimize disruptions and improve overall efficiency.

- **Example:** Financial institutions may use real-time streaming to quickly identify suspicious transactions and stop fraud, minimizing possible losses and safeguarding client accounts.

3. Competitive Advantage

Businesses can quickly respond to shifting consumer needs and market shifts thanks to real-time streaming, which gives them a competitive advantage. This agility allows organizations to stay ahead of competitors by continuously adapting their strategies based on the latest data. Responding promptly to market trends and customer preferences can lead to better market positioning and increased customer satisfaction.

- **Example:** Based on real-time sales data, retailers may constantly modify their inventory levels, guaranteeing that popular goods are always available and lowering the possibility of lost sales opportunities.

4. Predictive Analytics

Real-time streaming enhances predictive analytics by providing a continuous flow of data that improves the accuracy of forecasts and trend predictions. Organizations can leverage this capability to anticipate future needs and make data-driven decisions that drive innovation and efficiency. Real-time data supports timely interventions and strategic planning, leading to better outcomes and reduced uncertainties.

- **Example:** Healthcare providers can utilize real-time data to predict patient outcomes and tailor treatment plans accordingly, enabling more personalized and effective care.

5. Real-time Monitoring

Real-time streaming facilitates continuous monitoring of operations, which helps businesses maintain high levels of efficiency and minimize downtime. By constantly tracking key metrics and performance indicators, Organisations are able to promptly recognize and resolve possible problems before they become more serious. This ongoing oversight leads to more streamlined operations and improved overall performance.

- **Example:** Manufacturers can monitor machinery and production lines in real-time to detect and address any operational issues or maintenance needs, preventing costly disruptions and ensuring consistent product quality.

Stream analytics, also known as Real-time stream analytics, is the Real-time or near-time processing of data to predict future trends. In summary, event stream processing allows continuous data to be queried or analyze while responding to actual events in short intervals (usually milliseconds).

There are many time-tracking tools, such as:

1. **Apache Kafka:** Apache Kafka is a distributed streaming platform that creates streaming apps and real-time data pipelines. Kafka provides fault tolerance, low latency, and high throughput, and it can manage massive data streams. Kafka is a great solution for real-time data processing and analysis since it can interface with a variety of data sources and sink systems.

2. **Amazon Kinesis:** A fully managed real-time streaming data platform, Amazon Kinesis can handle data of any size and format. Real-time analytics, data exploration, and machine learning use cases may be supported by Kinesis, which has sub-second processing latency. Kinesis integrates with other AWS services and offers various security and compliance features.

3. **Apache Flink:** A free and open-source data flow analytics tool, Apache Flink facilitates the management of product and performance flows and aids in the computation of both restricted

and limitless data. Understanding, analyzing, and dividing stream data from many sources into different streams are all made possible by Flink. The interface of Flink is simpler to use and requires less training. Cluster management systems like YARN, Hadoop, and Kubernetes are accessible. Additionally, millions of events may be processed using Flink in milliseconds.

4. **Apache Hadoop:** A distributed database and MapReduce engine for storing and analyzing massive volumes of data are features of Hadoop, which is also free source. Although Hadoop is older and not as fast as Spark, many companies adopting it will keep going because something better is coming.

5. **Google Cloud Dataflow:** Google Cloud Dataflow is a fully managed batch and real-time data processing service that can process data from several sink systems and sources. An integrated programming model for batch and streaming data processing, Apache Beam, is supported by Dataflow. Auto-scaling, parallel processing, and connection with additional Google Cloud services are all provided by Dataflow.

6. **Apache Spark Streaming:** A real-time data processing framework based on Apache Spark is called Apache Spark Streaming. Spark streaming provides fault tolerance, high throughput, and low latency while processing data streams in micro-batch intervals. Spark Streaming can interface with other Spark components and supports a variety of data sources and sink systems.

7. **Azure Stream Analytics:** Data streams from a variety of sources, including mobile devices, can be produced via Microsoft Azure Stream Analytics products' real-time streaming architecture. Furthermore, data flow technologies like Cosmo DB and Azure Stream Analytics can read data streams, analyze previous data operations, and more.

8. **IBM Stream Analytics:** Customized, real-time streaming applications may be made with the aid of IBM Stream Analytics. This is powered by IBM streams, which allow users to safely

analyze, assimilate, or correlate data from several data sources in real time (Yadranjiaghdam et al., 2017).

Real-time analytics tools serve various purposes and offer numerous benefits for businesses.

Faster Decision-making: Businesses may analyze data in real-time or almost real-time with real-time analytics solutions, which helps them make choices fast as new information becomes available. This enables companies to react swiftly to opportunities, client demands, and changes.

- **Operational Efficiency:** Through real-time process monitoring and optimisation, it may assist companies in increasing operational efficiency. For example, they can detect and address bottlenecks, anomalies, and errors before they cause significant downtime or quality issues.
- **Customer Insights:** Customer satisfaction is a top priority for any business. It can give businesses Real-time insights into customers' behaviour, preferences, and needs. This enables businesses to personalize their offerings, improve customer engagement, and increase customer loyalty.
- **Risk Management:** Businesses may reduce risks by using real-time analytics solutions to monitor and detect threats in real time. For example, they can detect fraud, security breaches, and breaches before they cause serious harm.
- **Competitive Advantage:** This tool can give businesses a competitive edge by enabling them to make faster, more informed decisions than their competitors. This can help businesses to seize market opportunities, differentiate themselves from competitors, and improve their bottom line.
- **Innovation:** Real-time analytics tools can fuel innovation by enabling businesses to experiment, test, and iterate their products and services in real time. This enables companies to swiftly adjust to shifting consumer demands, market dynamics, and new technological advancements.

3.5 Designing Low-Latency Systems

In system design, latency is the amount of time it takes for a system to complete a job or react to a request. It is the interval of time between starting a task and getting a result. Latency in computing may happen in a number of ways, including hardware reaction times, data processing, and network connectivity.

Figure 3.7: Network Latency

Source: - *(GeeksforGeeks, 2024b)*

- The time lag between an action and its matching response is known as latency.
- Different units, such as seconds, milliseconds, and nanoseconds, can be used to measure it, depending on the system and application.

The client-server distance, data transmission speed, and network congestion are some of the variables that might affect latency in network systems. The system architecture, resource availability, and algorithm efficiency can all have an impact on data processing.

Minimising the lag or delay between the start of a request or operation and the anticipated answer or result is known as low latency. It's a crucial system design parameter, especially for real-time applications where prompt reaction or feedback is crucial. Low latency is crucial for system design because:

- **Enhanced User Experience:** Faster reaction times for consumers are guaranteed by low latency, which makes using apps more seamless and interesting. Low latency, for instance, is essential for online gaming because it gives players real-time responsiveness, which is essential for both enjoyment and competition.

- **Improved Efficiency:** Tasks are finished faster when latency is reduced, which enables systems to manage more requests or processes in the same amount of time. This results in increased system throughput and overall efficiency.

- **Competitive Advantage:** Low latency systems can offer a competitive advantage in sectors like banking, where snap choices can have a big impact. In order to take advantage of market opportunities before rivals do, traders depend on quick data processing and execution.

- **Real-time Data Processing:** Applications that need to handle data in real time, such as telephony or video streaming, depend on minimal latency to guarantee that information is delivered on time. A bad user experience might arise from buffering, delays, or lost connections caused by high latency.

- **Scalability:** Because low latency systems can manage higher loads without compromising responsiveness, they are frequently more scalable. Applications undergoing fast expansion or demand swings require this scalability.

- **Customer Satisfaction:** Low latency offers rapid access to material, lowers annoyance, and boosts user engagement for services like social media platforms and e-commerce, all of which improve consumer happiness.

Design Principles for Low Latency

Several concepts and tactics must be applied throughout a system's levels in order to design for low latency. These are some essential design guidelines for creating systems with minimal latency:

- **Minimize Round-Trips:** Instead of performing several separate requests, combine requests or transfer data in bulk to cut down on the number of round-trips between the client and server.

- **Optimize Network Communication:** Reduce network latency by using strategies including connection pooling, compression, and protocol optimisation. By utilising edge computing or Content Delivery Networks (CDNs), latency may be further decreased by minimizing the physical distance between users and servers.

- **Efficient Data Storage and Retrieval:** Use effective data storage techniques, such in-memory databases, caching frequently accessed data, or streamlining database queries to speed up retrieval times.

- **Parallelization and Asynchronous Processing:** To make better use of system resources and cut down on processing time overall, divide tasks into smaller components and carry them out asynchronously or in parallel.

- **Optimized Algorithms and Data Structures:** Select data structures and algorithms that put efficiency and speed first. For quick data access and retrieval, use data structures like hash tables or trees. For processing jobs, use algorithms with minimal temporal complexity.

- **Hardware Optimization:** To speed up processing and access times, spend money on high-performance hardware elements like CPUs, memory, and storage devices. If there are jobs that would benefit from parallel processing, use specialised hardware accelerators like GPUs.

- **Load Balancing and Scaling:** To avoid overwhelming any one component, use load balancers to divide incoming traffic evenly across several servers. Use auto-scaling techniques to dynamically modify resources in response to demand in order to keep latency low during periods of high load.

Low-latency systems that provide quick and responsive user experiences for a variety of applications and use cases may be developed by engineers by adhering to these design principles and consistently improving system architecture and implementation.

3.6 Use Cases for Real-Time Analytics

The Real-time analytics has made it possible for businesses to gather data in real time from operational infrastructure, equipment, and user interactions. They may now take immediate action on data as soon as it enters their systems. This may provide firms a competitive advantage by providing a wide range of use cases across several sectors, such as identifying financial fraud, accelerating the delivery of items in the supply chain, and improving factory inventory management.

Real-Time Analytics for Supply Chain

Through the use of real-time analytics to solve supply chain efficiency can be beneficial. In the UK, these inefficiencies nearly cost the country $2 billion. The supply chain business has a complex ecosystem since it involves a number of offline and online channels and players, including manufacturers and suppliers.

The goal of supply chain management is always to increase efficiency, speed, and cost savings; however, one of the challenges is the absence of real-time integration between both internal and external stakeholders. The problem of equipment failure also exists; a machine or piece of equipment is always susceptible to malfunctioning at a crucial moment. Finally, because batch data might be hours (or days) old, supply and demand data isn't always accurate when processed in batches.

Real-time analytics has changed the conversation from just automating procedures to integrating data in real-time and using it to improve decision-making. Real-time data streams may now be seen to improve supply chain management and demand and supply planning. Perhaps this explains why almost 66% of supply chain executives believe that analytics will play a crucial role in their operations going forward (Lechler et al., 2019).

Figure 3.8: Real-Time Analytics in Supply Chain Management

Source: - *(Kutay, 2024)*

Real-Time Analytics for Finance

The banking sector is one of the few that can employ real-time analytics to its full potential. This is due to the fact that it is synonymous with vast volumes of data, high volatility, and the requirement for real-time detection of intricate patterns. The capacity to correlate, analyze, and take action on financial data, such as transactional data, corporate updates, market prices, and trading data, is provided by real-time analytics. Every millisecond, this data comes in vast numbers from several sources, and banks and financial institutions must respond swiftly on it.

Real-Time Analytics for Manufacturing

Advanced analytics is "important" to 72% of factory executives, per a BCG poll. Even still, just 17% of them have found it to be "satisfactory" in terms of value. A clever application of real-time analytics may increase your operating efficiency, and there is much space for improvement.

Inventory management is one industrial process that real-time analytics can help you continually monitor, manage, and improve. Additionally, it lets you see how your production facility is operating in real time and can alert you to bottlenecks. This information may be gathered from equipment, sensors, CRMs, ERPs, and extra cameras that are placed throughout the building.

Figure 3.9: Real-Time Analytics in Manufacturing

Source: - *(Kutay, 2024)*

Personalization & Experience

One important use case for real-time analytics is the analysis of user behaviour to deliver personalized experiences. A consumer could be considering a product they previously bought, for instance. By offering this consumer a customized experience, you may raise the possibility that they will visit your shop again.

The so-called "next best offer" study is a typical illustration of predictive analytics. This method makes use of real-time analytics to present the user with a suitable offer based on their past interactions and behaviour. Giving a consumer of an online business relevant items depending on what they have viewed is a very basic example. Since the client is actively using the website, this analysis needs to take place in real time. The "personalized shopping experience" is a more comprehensive illustration. This method makes use of real-time analytics to give the user tailored offers according to their past interactions and behaviour. As you may have observed, this analysis is carried out in real time by merging transactional data—the browsing session you are now in—with historical data—the things you have previously purchased.

Fraud & Error Prevention

One important application of real-time analytics is the detection of fraudulent activity. For instance, it is possible to identify and prohibit questionable credit card transactions instantly. Although fraud may be detected using traditional analytical tools, these systems are excessively sluggish; it might take hours to analyze and analyze the data. Fraud may be immediately detected with real-time analytics.

The Anomaly detection, another name for the same concept, can be used to identify or avoid clerical mistakes. For instance, a vendor at an online store may be changing the costs of a range of goods. Inaccurate pricing and clerical mistakes might result from this. The solution is automated anomaly detection! The price can be fixed and the seller alerted if an anomaly is found.

Process Optimization

One of the objectives of digital transformation has been to optimize current procedures. However, digitizing a business process won't always make it better. Data collection and analysis are made possible by digitalisation, which yields insights that may be put to use. However, how can an ongoing process be optimised? You may, however, examine it in real time and make the required modifications. The procedure should, of course, permit modifications in real time.

One process that may be optimised in real time is production planning. You may automatically modify the production plan to satisfy client demands by integrating real-time analytics with the production planning process. Real-time demand tracking will enable you to build a more flexible supply chain.

The automation of repetitive human jobs is another facet of process optimisation. A contact centre representative, for instance, frequently responds to the same enquiries repeatedly. Many of the discussions with consumers may be automated with chatbots. To increase the chatbot's accuracy and efficiency, real-time analysis of consumer interactions with humans and chatbots is possible.

Pre-emptive Maintenance

In many businesses, preemptive maintenance may save maintenance expenses and downtime. For instance, a manufacturing business could be operating an equipment that is exhibiting symptoms of malfunction. You can address this issue right away if you can identify it. One important use for real-time analytics is this.

Application monitoring is an additional use case. System logs may be gathered and examined to identify faults instantly. This frequently enables you to fix the issue before it results in downtime. For instance, a web application's decreased performance may be observed. Before the website goes down, you can fix the issue and then launch the updated version.

3.7 Challenges in Implementing Real-Time Solutions

1. High Volume and Velocity

Challenge: Handling the enormous amount and speed of incoming data is one of the main difficulties in real-time data processing. Large volumes of data that enter continually at high rates, sometimes from several sources, must be handled by streaming data systems.

Solution:

- **Distributed Systems:** High-volume data streams may be managed with the use of distributed data processing frameworks like Apache Kafka and Apache Flink. By distributing data among several nodes, these systems allow for horizontal scalability and increase throughput.

- **Partitioning and Sharding:** Data streams can be partitioned according to time frames or keys to assist distribute the burden across several processing nodes. Data can be further dispersed across several databases or storage systems by sharding.

2. Low Latency Requirements

Challenge: In order to guarantee that data processing and insights are given in milliseconds or seconds, real-time applications frequently need exceptionally low latency. System performance, network overhead, and processing delays might make it difficult to achieve low latency.

Solution:

- **In-Memory Processing:** Latency may be greatly decreased by eliminating disc I/O operations by using in-memory processing frameworks like Apache Spark Streaming and Apache Flink. Faster execution speeds result from these frameworks' in-memory data processing.
- **Optimized Data Pathways:** Processing delays can be decreased by minimizing data transformations and using optimised data paths. Overall latency can be decreased by simplifying data pipelines and cutting out intermediary processes.

3. Data Consistency and Accuracy

Challenge: It can be challenging to guarantee data correctness and consistency in real-time processing, particularly when working with data from several sources. Inaccurate insights and judgements might result from inconsistent data.

Solution:

- **Exactly-Once Semantics:** Duplication and inconsistencies are avoided by implementing precisely-once semantics, which guarantees that every piece of data is handled exactly once. Surety of exactly-once processing is supported by technologies like as Apache Kafka and Apache Flink.

- **Data Validation and Cleansing:** Data accuracy may be preserved by putting in place real-time data validation and cleaning procedures. Problems like missing numbers or anomalies may be resolved by applying checks and filters to the data as it passes through the pipeline.

4. Fault Tolerance and Reliability

Challenge: Resilient systems that can recover from interruptions without losing data are essential for real-time data processing. Ensuring dependability and fault tolerance is essential for sustaining continuous data processing.

Solution:

- **Replication and Redundancy:** Fault tolerance may be improved by applying redundancy and data replication strategies. For instance, the replication function of Apache Kafka makes sure that data is replicated across several brokers so that errors may be recovered from.
- **Checkpointing and Recovery:** In the event of a failure, putting checkpointing and recovery methods in place aids in starting processing from a known state. For instance, Apache Flink allows checkpointing to regularly store the state of streaming applications.

5. Complex Event Processing (CEP)

Challenge: Real-time data processing frequently entails spotting and reacting to intricate patterns and occurrences, such spotting system irregularities or fraudulent transactions. Complex event processing can be difficult to implement since it requires advanced pattern recognition.

Solution:

- **Event Processing Engines:** Complex event processing may be made easier by using specialised event processing engines like Esper or Apache Flink. Rule-based processing, event correlation, and pattern matching are all supported by these engines by default.

- **Custom Rules and Algorithms:** Event detection skills may be improved by creating unique rules and algorithms suited to certain use situations. Deeper insights can be obtained by using machine learning models for predictive analytics or anomaly detection.

6. Integration with Existing Systems

Challenge: Planning is necessary since integrating real-time data processing technologies with legacy systems, databases, and data warehouses can be challenging.

Solution:

- **APIs and Connectors:** Data interchange and interoperability may be made simpler by integrating with current systems via APIs and connectors. Numerous frameworks for real-time data processing include connectors for well-known databases and data warehouses.

- **Data Transformation and Enrichment:** Integration may be facilitated by putting data transformation and enrichment procedures into place, which make sure the data is compatible with systems downstream. Tools for dynamic data transformation and enrichment are frequently included in real-time data processing systems.

7. Scalability

Challenge: It is essential to scale real-time data processing systems as processing needs and data quantities increase. Both managing higher data loads and making sure the system can support an increasing number of users or applications are components of scalability.

Solution:

- **Horizontal Scaling:** Scalability may be enhanced by designing systems for horizontal scaling, in which more nodes or instances are added to accommodate growing loads. Horizontal scalability is supported by distributed frameworks like as Apache Kafka and Apache Flink, which divide data and processing duties among several nodes.

- **Elastic Cloud Services:** Using elastic cloud services, such Google Cloud Dataflow or AWS Kinesis, enables dynamic scalability in response to demand. These services provide effective management of fluctuating data volumes by automatically adjusting resources based on workload.

8. Security and Privacy

Challenge: Strong security and privacy safeguards are necessary when handling sensitive data in real-time to guard against data breaches and illegal access.

Solution:

- **Encryption:** Encrypting data both in transit and at rest guarantees that private data is shielded from unwanted access. Encryption protocols are supported by many real-time data processing frameworks to protect data.
- **Access Controls:** Data processing systems can be protected by putting in place authentication procedures and fine-grained access controls. By limiting access to authorised users, multi-factor authentication (MFA) and role-based access control (RBAC) can improve security.

3.8 Data Ingestion and Processing Pipelines

Data ingestion is the procedure for gathering and importing data files into a database for processing, analysis, and storage from several sources. Data ingestion aims to prepare data for usage inside the company by cleaning and storing it in a single repository that is consistent and easily accessible.

An enterprise system may use data ingestion tools to import data from dozens or even hundreds of individual sources, including:

- Internal databases
- External databases
- SaaS applications
- CRM systems
- Internet of Things sensors
- Social media

Types of Data Ingestion:

1. **Batch Ingestion:** Transfers large volumes of data at scheduled intervals. It is cost-effective and well-suited for periodic updates.
2. **Real-Time Streaming:** Continuously streams data into the system, enabling immediate access and analysis. This is critical for applications like IoT monitoring and live dashboards.

Table 3.2: Types of Data Ingestion

Feature	Batch Ingestion	Real-Time Ingestion
Timeliness	Scheduled updates	Immediate updates
Resource Use	Lower computational need	Higher computational need
Use Cases	Historical reporting	Live dashboards, IoT

Source: - *(Dutta, 2024)*

Data ingestion can occur in batches or in real-time streams. Batch ingestion involves transferring large chunks of data at regular intervals. With streaming ingestion, data is continuously transferred into the system. Typically, real-time streaming ingestion delivers more timely data into the system faster than batch ingestion does.

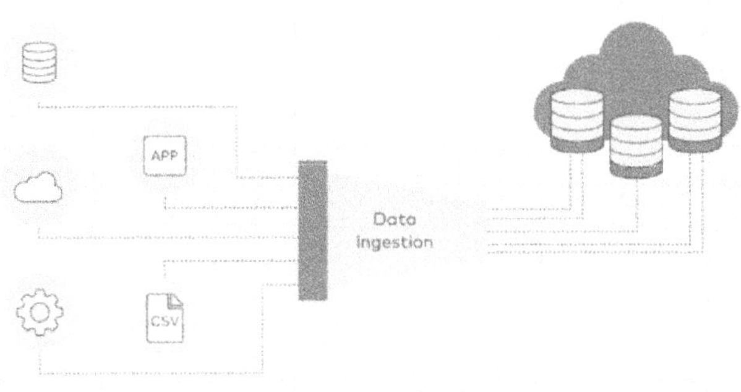

Figure 3.10: Data Ingestion

Source: - *(Dutta, 2024)*

A data ingestion pipeline is a system that connects multiple data sources to centralized storage, such as a data warehouse or lake. It ensures the smooth movement of data, enabling organizations to structure and organize it for analysis.

A well-designed data ingestion pipeline architecture ensures that data flows seamlessly from its sources to its destination. It must handle diverse formats, support scalability, and incorporate monitoring to maintain data quality.

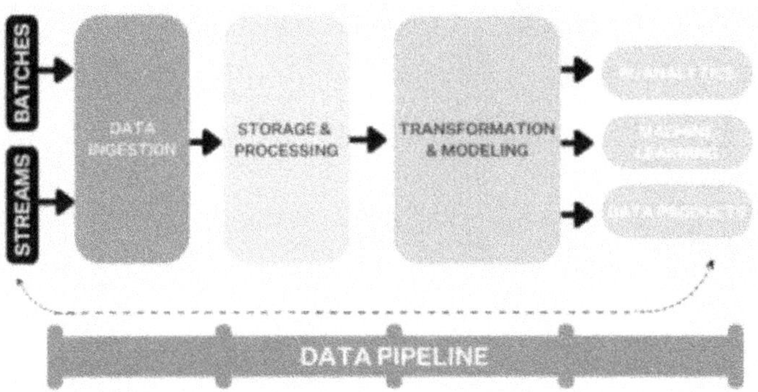

Figure 3.11: Data Ingestion Pipeline Architecture

Source: - *(Dutta, 2024)*

Step-by-Step: How Data Ingestion Process Flows Work

Process Flow:

1. **Ingestion:** Data is collected from various sources.
2. **Processing:** Data is cleaned, validated, and transformed.
3. **Storage:** Data is centralized in repositories for further analysis.
4. **Analysis:** Data is accessed for insights, reporting, or machine learning applications.

A typical data pipeline has six key data ingestion layers

- **Data ingestion**: This layer accommodates either batched or streamed data from multiple sources.
- **Data storage and processing:** Here, the data is processed to determine the best destination for various analytics, and then stored in a centralized data lake or warehouse.
- **Data transformation and modeling:** Given that ingested data comes in diverse sizes and shapes, not all of it is formally structured (with IDC estimating that 80% of all data is unstructured), this layer transforms all data into a standard format for usability.
- **Data analysis:** This final layer is where users access the data to generate reports and analyses.

Organizations with different data needs may design their data pipelines differently. For instance, a company that only uses batch data may have a simpler ingestion layer. Similarly, a firm that ingests all data in a common format might not need the transformation layer. The data pipeline should be customized to the needs of each organization.

3.9 Chapter Conclusion

The chapter concludes that real-time analytics is now a key component of modern data- driven businesses as it provides the much-needed link between the raw data and business intelligence in real time. However, it brings about decision-making benefits such as customer satisfaction, operation enhancement, and decision-making enhancement, and implementation issues involving data velocity and data consistency come into play. Nonetheless, the state-of-the-art architectures, tools, and design principles suggest that these challenges become more tractable over time. In this concept of continued industrial digitalization, real-time analytics cannot be overlooked as an essential tool to help organizations stay relevant in a constantly shifting market environment.

Multiple Choice Questions (MCQs)

1. What distinguishes real-time analytics from batch processing?

- a. It processes data periodically
- b. It delivers insights within milliseconds
- c. It focuses only on historical data
- d. It does not require continuous data flow

2. Which architecture is commonly used in real-time data systems?

- a. Monolithic
- b. Lambda
- c. Waterfall
- d. Star schema

3. Which tool is widely used for real-time data streaming?

- a. Apache Hadoop
- b. Apache Kafka
- c. MongoDB
- d. Tableau

4. Real-time analytics in e-commerce helps improve:

- a. Batch reporting
- b. Customer experience
- c. Inventory prediction only
- d. Offline processing

5. What is a key challenge of real-time analytics systems?

a. High latency
b. Slow data ingestion
c. Handling high data velocity
d. Static data models

6. The Kappa architecture focuses on:

a. Real-time and batch processing
b. Real-time data processing only
c. Batch processing only
d. Data warehousing

7. What does the term "data velocity" refer to?

a. Data variety
b. The speed at which data is generated and processed
c. Data size
d. Data structure

8. Real-time fraud detection in finance is enabled by:

a. Historical batch processing
b. Predictive maintenance
c. Real-time data streaming
d. Data visualization

9. Which of the following tools is used for secure real-time data transmission?

a. Google Big Query
b. Apache Spark
c. Amazon Kinesis
d. Matplotlib

10. Real-time analytics enables organizations to:

a.   Make decisions based on past events only
b.   Respond quickly to ongoing events
c.   Reduce the need for data governance
d.   Replace machine learning models

Answer

1	2	3	4	5	6	7	8	9	10
b	b	b	b	c	b	b	c	c	b

Chapter 

ADVANCED TECHNIQUES IN DATA ENGINEERING

4.1 Chapter Overview

This Chapter focuses on the latest methods and tools to improve data management and processing in today's organizations. It starts with data modeling that outlines practices such as visualization, creation of a single source of the truth, and an iterative approach to building good data systems. The chapter also discusses ETL (Extract, Transform, Load) pipelines and their application with regard to the automation of data movement and transformation processes as well as considerations of data quality.

Moreover, it extends the difficulties of processing the semi-structured and unstructured data, which are suggested as the usage of the NoSQL databases, the tools of the distributed storage, and pre-processing. The synergy between data engineering and machine learning is presented in details, with the emphasis on the aspects of data pipeline management, feature extraction, on-line prediction, and model performance evaluation. Further topics are emerging orchestration tools like Apache Airflow and Dagster, and the question of data lineage and metadata, which are crucial to stability and compliance. This regards the chapter emphasizing the role of data engineering in using modern methods to improve processes, foster innovation and enhance the possibility of creating efficient large-scale data systems.

4.2 Data Modeling And Its Best Practices

As the primary step in software engineering, data modelling entails assessing all of the application's data dependencies, clearly describing (usually through visualizations) how the data will be used by the software, and defining data objects that will be kept in a database for future use.

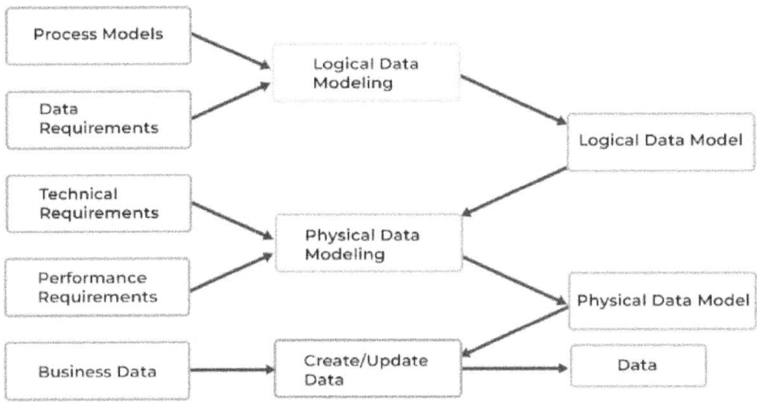

Figure 4.1: How Data Modelling Works

Source: - *(BasuMallick, 2022)*

The act of creating a visual depiction of all or a portion of an information system to demonstrate the connections between various data points and organisational structures is known as data modelling. The objective is to describe the many forms of data that are utilised and kept in the system, as well as how they are related, how they may be arranged and grouped, and what their formats and characteristics are.

Business needs are considered while creating data models. Before being incorporated into the design of a new system or modified during an iteration of an existing one, rules and requirements are created with input from business stakeholders.

Since a data model graphically depicts the relationships between data entities, their different qualities, and the nature of the data entities themselves, it is comparable to a flowchart.

Before any code is written, data management and analytics teams may use data models to identify errors in development plans and outline the data needs for apps. Alternatively, data models can be created by attempting to reverse engineer them out of existing systems(Carvalho et al., 2022).

When embarking on a data modeling project or task, one should remember the following best practices:

1. Design the data model for visualization

Enlightenment is unlikely to result by staring at infinite rows and columns of alphanumeric information. Many individuals are comfortable using user-friendly drag-and-drop screen interfaces to rapidly analyze and combine data tables or seeing graphical data visualizations that make it simple to identify any abnormalities.

You may make your data complete, error-free, and redundant-free by cleaning it with methods like these for data visualisation. They also assist in determining different sorts of data records that correspond to the same physical object so that you may convert them into standardised fields and formats to make it easier to combine different data sources.

2. Recognize the demands of the business and aim for relevant results

To help an organization function more effectively is the goal of data modeling. When seen from the point of view of a trained professional, the most significant challenge posed by data modeling is the precise capturing of business needs. Determining which data should be collected, stored, altered, and made accessible to users requires this.

You can get a complete picture of the needs by asking consumers and stakeholders what they want from the data. You should begin organizing your data with those objectives in mind. Starting to carefully structure your data sets with the requirements of users and stakeholders in mind is preferred.

3. Establish a single source of truth

Bring the raw data from all of your sources into your data warehouse or database. If you just use "ad-hoc" data extraction from the source, it might affect the flow of your data model. If you use all of the raw data stored in your centralised hub, you will have access to all of the historical data. Making calculations using logic on data that has been extracted straight from a source might backfire or even ruin your entire model. If something goes wrong in the process, it is also very hard to fix or maintain.

4. Start with simple data modeling and expand later

Data may quickly become complicated due to factors including size, type, structure, growth rate, and query language. When data models are initially kept small and basic, it is easier to address problems and take the appropriate actions. Once you are certain that your first models are accurate and meaningful, you may add more datasets, eliminating any discrepancies along the way. Finding a tool that is initially easy to use yet eventually supports very massive data models is a good idea. Additionally, it must to enable you to rapidly integrate several data sources from different physical locations.

5. Before moving on, double-check each step of your data modeling

Every task should be reviewed twice before moving on to the next phase, beginning with the data modelling priorities determined by the business needs. For example, selecting a primary key for a dataset guarantees that the value of the main key in each record may be used to uniquely identify each record in the dataset. Two datasets can be combined using the same technique to see if there is a one-to-one or one-to-many link between them and to avoid many-to-many interactions that lead to unmanageable or too complex data models.

6. Organize business queries according to dimensions, data, filters, and order

Understanding how these four variables could be utilised to clarify business problems is made easier with well-organized data sets. For instance, if a retail company has stores all around the world, it is possible to determine which ones did the best last year. The product and store location are the dimensions, the filter is "last 12 months," the facts are sets of historical sales data, and the order is "best five stores in declining order of sales. "You may support the study by determining the best sales performers for each quarter and correctly answering further business intelligence questions by meticulously organizing your data sets and utilising separate tables for dimensions and facts.

7. Perform computations beforehand to prevent disputes with end customers

It is crucial to have a single truth version that users may do business against. Although there may be disagreements about how the answer should be utilised, there should be no debate about the underlying data or the math that was used to get there. To determine the best and worst months, for example, a formula may be required to aggregate daily sales data into monthly values.

A company can avoid issues by including this calculation into its data modelling beforehand rather than requesting that everyone use their own calculators or spreadsheet software.

8. Search for a relationship rather than just a correlation

Instructions for using the modelled data are part of the data modelling process. Giving people the ability to access business analytics on their own is a big step, but it's just as crucial that they refrain from making mistakes. For example, that can be the case if we see that the sales of two unrelated items seem to increase and decrease in tandem. Do sales of one product influence sales of another, or do they fluctuate in tandem in response to external factors like the weather and the economy? In this case, a perplexing relationship and correlation may be pointed in the wrong way, which would deplete resources.

9. To carry out complex jobs, use contemporary tools and methods

Data sets can be prepared for analysis by programming before more intricate data modelling is carried out. But what if there was a tool or application that could manage these challenging tasks? People no longer need to learn many coding languages, which frees up your time to work on projects that will benefit your business. Extract, transform, and load tools are examples of specialised software that can help or automate all of the data extraction, transformation, and loading operations. Data modelling may even be done automatically, and several data sources can be combined using a drag-and-drop interface.

10. Enhanced data modeling for improved business results

Among other things, data modelling that helps users quickly discover answers to their business issues may improve the company's performance in the areas of effectiveness, yield, competence, and customer satisfaction. The use of technology to expedite the stages of examining data sets for answers to all enquiries, as well as in connection with tools, business goals, and organisational objectives, are essential elements. Setting data priority for different corporate operations is another aspect of it. Having fulfilled these scenarios will enable your organisation to more reliably predict the important benefits and productivity gains that data modelling will deliver.

11. Verify and test the application of your data analytics

As with any other built-and-implemented functionality, test your analytics' implementation. To determine whether the quantity and precision of the whole data collection are accurate, it should be checked. Think about whether your data is properly arranged and allows you to derive a crucial metric. You may also construct some queries to better understand how it will work and be used. We also suggest developing a range of projects to evaluate your implementation and execution.

4.3 Implementing ETL Pipelines

Extract, transform, and load is what **ETL** stands for. This group of procedures that transfers data from a source system to a destination system is known as ETL.

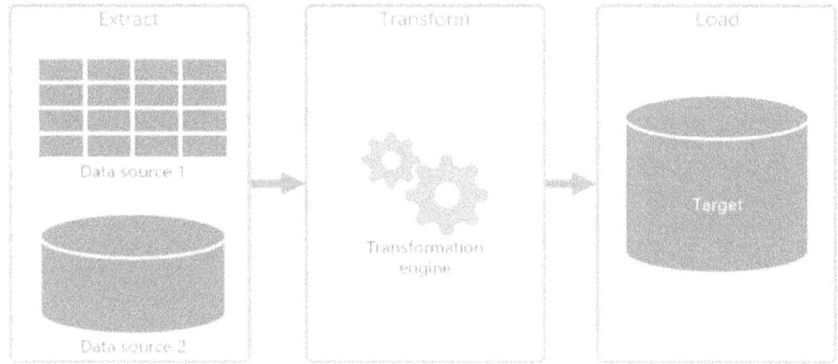

Figure 4.2: ETL Pipelines

Source: - *(Richman, 2024a)*

Transferring data within an organisation is crucial. A contemporary business usually has a lot of diverse data sources and varied objectives that need data to be used by different systems and in different ways.

They let companies to automatically transfer data between systems, incorporating data transformations as necessary.

Extract

Information and data must initially be extracted, or ingested, by the pipeline from one or more sources.

Data source systems often include:

- APIs
- Websites
- Data lakes
- SaaS applications
- Relational databases, or transactional databases

The precise pipeline design determines the technical specifics. The pipeline needs some kind of interface in order to interact as it is a different kind of data system from the source: An incorporation.

A vendor may sell an ETL tool to a business. In this instance, the tool often comes with integrations with many source systems. In other situations, the company's data engineers may construct the ETL pipeline itself, including the required source connectors.

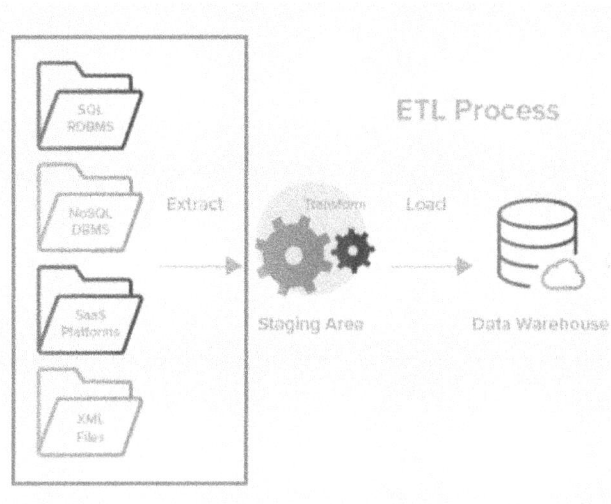

Figure 4.3: ETL Process

Source: - *(Richman, 2024a)*

Real-time pipelines may extract data from the source as soon as it is visible in the source. Additionally, they can employ a batch processing methodology, which picks up new changes at predetermined intervals.

Transform

The second phase in the ETL process is transformation. This stage converts the raw data that was taken from the source into data that is appropriate for the particular use case and destination.

For instance, your source data may be in CSV format, but your business intelligence tool may need JSON data that follows a specific structure. As the data moves through your ETL pipeline, the transformation phase may convert it from one format to another.

There are many different use cases; that is only one example. One thing, however, is nearly always true: data transformation is required at some stage of its lifespan. That change occurs practically instantly once the data is extracted in an ETL process.

What types of data processing happen during the transformation step?

Typically, they consist of basic procedures such as:

- Filtering
- Aggregation
- Data cleaning
- Feature extraction
- Re-shaping data to conform to a schema

There is a crucial reason behind these comparatively easy changes. They guarantee that the dataset is error-free, clean, and consistent whether it is transferred to storage or another application. It is prepared for use in achieving company objectives.

They are very easy to automate due to their simplicity. Members of the data team save time by doing this instead of doing a laborious activity.

On the other hand, extensive analysis and more intricate, exploratory data transformations might take place later in the data life cycle. This task is more complicated and typically calls for human involvement. However, you may get here more quickly by automating the simple tasks up front.

Load

Data loading into the destination system, also known as the target system, is the final stage of the ETL process. This is extraction's opposite.

A data warehouse or other kind of data repository intended to support data analysis is the destination in a common, everyday use case. Numerous analytical procedures can occur once in the warehouse, including: exploratory analytics, machine learning, business intelligence, and more.

But there is a vast range of potential destinations and application cases. In general, destination systems can consist of:

- Data warehouses
- SaaS applications
- Operational business systems
- Visualizations and dashboards

To load data, the pipeline must interface with the destination system, much like the source. Whether you utilize an ETL tool, what kind of tool you employ, and whether your ETL pipeline was constructed internally all affect the loading process's technical aspects.

After the data has been loaded, you may begin to observe the data flow's latency, or how long it takes for the ETL data to go from source to destination.

Depending on the pipeline, latency can range from hours to minutes or, in the case of real-time data streaming, milliseconds (Pogiatzis & Samakovitis, 2021)Transform, and Load (ETL.

4.4 Handling Semi-Structured and Unstructured Data

Semi structured and unstructured data are common in data engineering today with data coming from social media platforms and IoT devices. This kind of data is often processed using specific methods and tools because of their heterogeneity and free structure. This paper aims at analysing the various approaches and recommendations for dealing with such kinds of data efficiently.

Understanding Semi-Structured and Unstructured Data

Semi-structured data is at least partially ordered, but it uses tags or markers to delimit data fields, e.g., XML, JSON or NoSQL databases. Structured data on the other hand does not have a pre-defined model and includes text, images, videos, logs among others. Both of them threaten conventional disjointed record systems and require new approaches to storing, managing and analysing data.

- **Data Storage Solutions:** Selecting the right sort of storage is important. For semi-structured data, there are NoSQL data stores such as MongoDB and Cassandra where data have a flexible structure, and hence it is straightforward to incorporate data of different kinds. For unstructured data, distributed storages like HDFS or a service like AWS S3 or Google Cloud Storage as these are very scalable and inexpensive to manage large volumes of data.

- **Data Pre-processing Techniques:** Pre-processing includes undertaking activities on the data collected so that it can be in a form suitable for analysis. Semi-structured data require related tools, such as Python libraries (pandas and json) or the ETL (extract, transform, load) tools to handle and cleanse data. When dealing with unstructured data, pre-processing might involve text mining for textual data, image processing for graphical data and all sorts of feature extraction techniques to bring all data types to a similar form.

- **Data Integration and Transformation:** Blending of both semi-structured and unstructured data with structure data sources need considerable workflows. Apache Nifi, Talend and Apache Kafka are some of the tools which help in real time ingestion and transformation. Standardising schemas and meta tagging can assist with the input's compatibility issue and make the integration process faster.

- **Analytical Frameworks:** Sophisticated means of analysis and modeling are required for efficient data analysis. Semi- structured data can be queried using SQL like operations in tools like Apache Hive or Google Big Query. Open source tools for unstructured data includes; Machine learning algorithms and frameworks, TensorFlow, PyTorch and OpenCV. These frameworks can take image, sound, and text data and, in turn, model out significant patterns.

- **Handling Data Quality and Governance:** Keeping data accuracy is crucial since most of the data gathered is unstructured and semi-structured, which might include ample redundancy and noise. Databricks or Alation are the automated tools that enforce compliance and help in the data accuracy of Governance. Proper metadata management and cataloguing when used takes discoverability and usability to another level.

- **Scalability and Performance Optimization:** The technical architecture is divided into scalable architectures that are effective in handling semi-structured as well as unstructured data in large volumes. Some examples of distributed computing frameworks are Apache Spark or Kubernetes which help with parallel computation and resource scheduling. Implementation of caches and indexes allows for efficient and fast query preparation and often, even provide fast access to the data set (Kumar, 2021)leading to further growth in the variety of data. Initially, the prominent organization faces difficulties due to the exponential development of data. The benchmarking is required to allow data mining tools such as Rapid Miner and Big Data in the environment. Structured, unstructured, and semi-structured data are the types of Big Data. Structured data are used to develop a page by giving enough information. Unstructured data will not have a predefined model to arrange the data in a particular way. Semi-structured data has some structure, but it will not have any data model. The extensive data system is the more apt method in evaluating the benchmark. The proper research and analysis must be carried out to provide a standardized benchmark. The organization has to develop a clear-cut idea about the strategy to overcome the consequence effectively. The benchmark is usually classified into two types internal and external benchmarking. Internal benchmarking includes SWOT, performance metrics, functional values, a practice carried out, financial information, and external benchmarking, including collaborative benchmarking of process, product, corporate, strategic, and global. The primary purpose of the internal benchmark is to compare the performance of an organization internally. The external benchmark gives information about the organization's performance with its competitor companies in this industrial world. The primary big bench model for the product retailer includes a synthetic data maker with variety, speed, and extensive data systems, which contains structured and unstructured data. The significant bench workload is performed with a set of queries. The queries occupy various departments of comprehensive data analysis based on a business perspective. Data sources, processing types of queries, and analytic techniques are included in designing the queries from a technical perspective. Transaction Processing Performance Council decision support (TPC-DS.

4.5 Workflow Orchestration Tools

Workflow orchestration is a method of coordinating and managing complex processes across multiple automated tasks and systems within an organization. Unlike simple task automation that executes repetitive actions, workflow orchestration involves overseeing the logical flow of these tasks, ensuring they interact seamlessly and efficiently to complete broader business processes. It acts as the conductor in an orchestra, directing various automated activities to work in harmony and achieve specific business objectives.

This strategic management layer is crucial for businesses that rely on diverse systems and applications, as it integrates and synchronizes workflows to prevent silos, reduce redundancies, and ensure that data flows smoothly across all points of the process. By doing so, it facilitates more sophisticated, scalable, and adaptable business operations.

Workflow orchestration significantly enhances business operations by automating and synchronizing complex processes across various systems. By strategically allocating resources, cutting down on execution times, and minimizing mistakes, this task management improves productivity and efficiency. As businesses grow, workflow orchestration enables them to scale operations effectively without proportionate increases in complexity or costs. This results in improved service delivery, higher customer satisfaction, and ultimately, a stronger bottom line, making businesses more agile and competitive in the marketplace.

There are various Workflow Orchestration Tools, which are as follows:

- **Nected**

Nected is a potent Workflow Orchestration solution made to automate, optimize, and simplify intricate business procedures. Nected gives businesses the ability to manage tasks, rules, and data flows across systems with ease by emphasizing the integration of decision-making skills into workflows. It is perfect for companies looking to increase productivity, save operating expenses, and improve overall process agility because of its intuitive interface and sophisticated orchestration features.

Nected stands out as a top Workflow Orchestration tool due to its ability to integrate decision-making capabilities within automated workflows. Its no-code/low-code interface makes it accessible for a wide range of users, while its powerful rule engine and seamless data integration ensure intelligent, data-driven processes. With Nected, organizations can enhance operational efficiency, reduce costs, and adapt quickly to evolving business needs. Whether it's automating routine tasks, managing complex workflows, or enabling smarter decisions at every step, Nected offers the tools and flexibility needed to drive business success and transformation.

- **Google Workflows**

Google Workflows is a robust workflow orchestration service provided by Google Cloud, designed to sequence and automate tasks across Google services. It is particularly effective for businesses heavily embedded in the Google Cloud ecosystem, looking to automate and optimize their operations without extensive coding.

- **Microsoft Power Automate**

An essential component of the Microsoft ecosystem, Microsoft Power Automate (previously known as Microsoft Flow) helps companies automate processes and boost productivity across a range of apps and services. It is particularly powerful for organizations that utilize Microsoft 365 services extensively.

- **AWS Step Functions**

AWS Step Functions is a powerful service offered by Amazon Web Services that orchestrates workflows for applications. It enables developers to create and implement workflows that automate tasks across a range of AWS services, including Amazon SNS, Amazon DynamoDB, and AWS Lambda. This tool is particularly effective for businesses deeply integrated into the AWS ecosystem, providing a reliable and scalable solution to manage complex workflows.

- **Dagster**

Dagster is an open-source data orchestrator for creating, managing, and monitoring workflows and data pipelines. Dagster aims to give data scientists and engineers a productive development environment that facilitates the definition, testing, and implementation of data processes.

- **Argo**

An open-source container-native workflow engine called Argo is used to coordinate many tasks on Kubernetes. It is specifically designed to handle complex dependencies with advanced scheduling capabilities, making it highly suitable for jobs that need to run in large-scale distributed environments.

- **Camunda**

A robust workflow orchestration tool, Camunda was created to assist businesses in automating, visualizing, and streamlining their operations. It is particularly well-suited for enterprises that require robust workflow and decision automation. Camunda provides both an open-source core and proprietary features that cater to complex process orchestration needs.

- **Apache Airflow**

An very flexible open-source solution for process scheduling and monitoring is Apache Airflow. It was created by Airbnb and uses directed acyclic graphs (DAGs) to programmatically author, schedule, and monitor processes. Airflow is widely recognized for its robust community, extensive customization capabilities, and its ability to handle complex data pipelines.

- **Salesforce Orchestrator**

Salesforce Orchestrator is a low-code workflow orchestration tool designed to automate and manage complex, multi-user, multi-step business processes within the Salesforce ecosystem. By synchronizing processes across several teams and departments, it helps businesses to increase productivity, streamline operations, and improve customer experiences.

- **Temporal**

Temporal is an advanced workflow orchestration platform designed to simplify the management of complex, long-running, and distributed workflows. It supports durable execution and is highly scalable, making it suitable for a variety of business and technical use cases. Temporal provides an open-source core with additional proprietary features and services for enhanced functionality.

- **Active pieces**

Active pieces is an open-source, no-code business automation tool designed to streamline workflows and enhance productivity. It offers a robust set of features that cater to various business needs, from marketing and sales to IT operations and customer support. Active pieces allows users to automate tasks and integrate multiple applications seamlessly (Corodescu et al., 2021).

4.6 Integrating Machine Learning with Data Engineering

The combination of data engineering with machine learning (ML) is changing the ways organizations work with data. When incorporating the best practices of data engineering with data science and machine learning, organizations can gain insights and make decisions that, in turn, will help them improve their processes. This integration needs proper organization in order to have proper data and model output exchange.

1. Building a Data Foundation for Machine Learning

Data architecture is important when combining ML with data engineering. This involves the method of accumulating, manipulating and archiving data in a format that is suitable for the development and application of ML models. The data engineering pipelines should have to some extent data quality, data consistency and accessibility. Framework such as Apache Spark, Apache Kafka, and cloud-based solutions to offer the architecture for handling Big data and Big speed data for ML.

2. Feature Engineering and Data Transformation

Feature engineering plays the role of a link between data engineering and ML. Feature extraction mostly entails understanding the raw data and converting them into usable features which may only be achieved by a combination of domain knowledge and technical knowledge. ETL procedures and frameworks including pandas, PySpark, or Scikit-learn are used by data engineers to develop features to improve the performance of models. To avoid this, feature engineering is made more automated for example through feature stores to enable reusability across multiple projects.

3. Model Training Integration

The process of model training within data pipelines typically requires the real-time automation of data acquisition and data preparation. Such frameworks as TensorFlow Extended (TFX) and MLflow eliminate the need for such pipelines because they link data pipelines with ML processes. With such tools, data engineers and ML practitioners are able to work through models at a faster pace and be assured that the training data sets are current.

4. Real-Time Data and Model Serving

Integration between data pipelines and ML models allows data to flow through the pipeline and make predictions as soon as the data arrives. Predictions may be made straight from the streaming data when the models are trained as machine learning models using tools like TensorFlow Serving, FastAPI, or Kubernetes. For effective inference and real-time decision-making, real-time stream processing is accomplished using Apache Kafka and Flink.

5. Monitoring and Feedback Loops

It is crucial to control data flow and ML models in order to keep both metrics at the desired level. Data engineers can put into practice monitoring systems that would observe and report on the quality of the data as well as identify irregularities. At the same time, there is a possibility to use ML monitoring tools such as Prometheus and Grafana for tracking the model's shift and effectiveness over time. The use of feedback loops in Machine learning involving recollecting new data to train the models guarantees that the predictions are still valid in the current conditions.

6. Scalability and Resource Management

Another principle of machine learning together with data engineering is scalability. Apache Spark, Kubernetes, or any other cloud-based platforms like AWS, Azure, or GCP contain all the necessary resources for dealing with large-scale data and complex models. It is confirmed that resource management and load balancing are effective in maintaining high productivity of the system when hit with high loads (Jayabalan, 2024).

4.7 Monitoring and Debugging Data Pipelines

A crucial procedure that guarantees the smooth transfer of data from source to destination for in-depth analysis is data pipeline monitoring. Organisations can preserve data quality, reduce the possibility of user mistake, get rid of OCR data inaccuracies, and prevent other issues that might affect important analytical procedures by actively monitoring data pipelines.

In a time when strict adherence to data governance standards like the CCPA and GDPR is required, it becomes strategically necessary to monitor the quality of data as it moves between different locations in order to prevent significant fines from the government. Additionally, proactive pipeline monitoring strengthens security protocols by limiting unwanted access to private information and enhancing the organization's defenses against any breaches.

You'll discover a variety of useful techniques and resources for monitoring data pipelines below. Investing in monitoring and data visualisation tools, utilising ETL/ELT for pipelines, and routinely evaluating pipelines are some of the most popular approaches.

Setting precise goals that are in line with the requirements of the company is essential before starting data pipeline monitoring. For example, the requirement to monitor data quality may be prompted by adherence to data governance standards such as the CCPA and GDPR. It's critical to monitor data quality as it travels from one place to another since noncompliance with these guidelines may result in costly fines from the government.

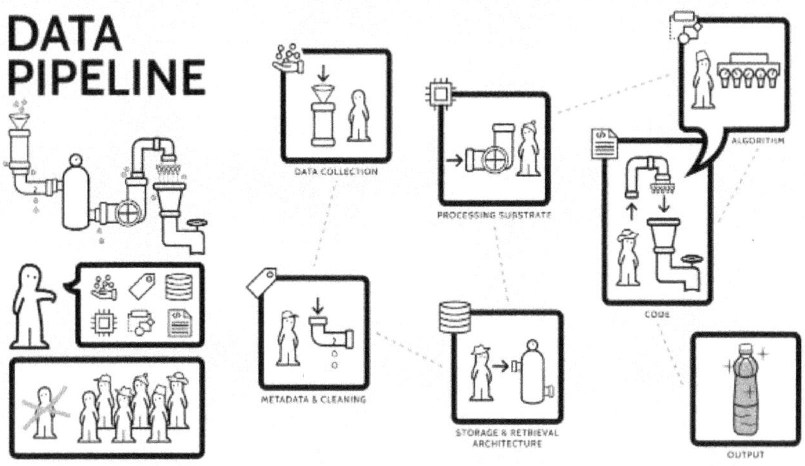

Figure 4.4: Data Pipeline Monitoring

Source: - *(Dearmer, 2023)*

Organisations must also identify relevant Key Performance Indicators (KPIs) to evaluate the success of their data pipeline monitoring programs.

These KPIs can include:

- Data latency, which tracks the duration of data flow via your pipelines
- Availability, which tracks how long your pipelines are operating properly
- Utilisation, which keeps track of how much CPU and disc space your pipelines are using at any one time.

The objectives for data pipeline monitoring, the monitoring tools you employ, and the kind of pipelines that have been created—ETL/ELT pipelines, event-driven pipelines, streaming pipelines, etc.—will all influence the KPIs that have been selected.

Organisations may efficiently monitor and enhance their data integration processes by coordinating goals with certain KPIs.

Selecting the appropriate monitoring tools for data pipeline monitoring should come after you have established your goals and KPIs. In order to find any issues and make sure everything is operating as it should, this software program collects data about the general health and functioning of the data pipelines in your company. Monitoring instruments consist of the following:

- **Logging frameworks**

Logging frameworks gather log data from target systems and data sources, among other components engaged in data pipelines. These tools capture various events that might affect pipelines, such user problems, and provide logs for particular data pipeline processes. Real-time log data collection by the top logging frameworks enables you to spot mistakes as they occur and determine their underlying cause.

- **Observability tools**

Although they also gather data from many pipeline components, observability tools offer a completer and more all-encompassing picture of the pipeline, including the locations of data storage and processing methods. Additionally, these technologies offer information on how to optimise business intelligence workloads through the usage of data. Artificial intelligence, machine learning, predictive analytics, and other techniques are used by the top observability technologies to produce precise data insights that give your team knowledge about pipelines.

- **Data visualization tools**

Data visualisation tools let you get insights into your data pipelines by displaying various metrics and KPIs via observability tools and logging frameworks. A data visualisation tool may help you better understand your data pipeline activities by presenting data sets as graphs, charts, reports, and dashboards.

Monitoring a data pipeline effectively is a continuous activity rather than a one-time event. Monitoring is useful for more than just building pipelines. To adhere to data governance regulations, prevent bottlenecks, produce useful business insight, and enhance data integration workflows, you should keep an eye on data as it moves between locations. Tracking your pipelines as your company grows is another aspect of continuous monitoring. You need pipelines that can effectively transfer the growing amount of big data in your organisation to the appropriate destination.

Documenting the performance of data pipelines over time and sharing the findings with your data engineering team is one of the greatest methods to assess them. You may build even more effective pipelines in the future by learning how pipelines automate data integration and transfer data between a source and a destination, such a data warehouse or data lake. To make sure data goes to the right place without any problems, data pipelines should be tested on a regular basis.

4.8 Data Lineage and Metadata Management

Capturing, arranging, and preserving metadata—which offers crucial context and details about the data itself—is the task of metadata management. Data management is what characterizes other data.

To put it briefly, it is the process of managing information about your data. In a database, data warehouse, or data lake, metadata is usually utilised for information management, discovery, and comprehension.

<u>**Types of metadata management**</u>

There are several kinds of metadata, including as:

1. **Structural metadata:** This helps to clarify the links between various data sets or pieces and explains how data is organized.
2. **Descriptive metadata:** This gives consumers information about unique data instances, making it possible for them to locate certain data items inside a larger data collection.
3. **Administrative metadata:** This comprises technical details that aid in resource management, like the creation date, method, and availability of data.

Data lineage is yet another crucial idea in data governance. It offers details on the following:

- Origins of data set
- How does it move over time?
- What happens between the time of its genesis and the present?

Data lineage may be used to trace mistakes back to their origin and determine how modifications to upstream data may affect downstream operations. Additionally, it facilitates adherence to a number of rules.

Types of data lineage

There are several forms of data lineage, including:

- **Forward lineage:** It facilitates the visualisation of data flow from source to destination.
- **Backward lineage**: It aids in identifying the data's source.
- **Horizontal lineage**: It aids in demonstrating how data moves across systems and procedures.

Interrelation between metadata management and data lineage

Data lineage is essentially an essential component of the metadata. It is a crucial component of "descriptive metadata," which gives details about the history of the material.

Who has accessed the data, what modifications have been made, when these changes were made, and why are some examples of the information that may be recorded?

Understanding and preserving data provenance may be greatly aided by having a strong metadata management system.

It will provide a clear view of:

- Where data comes from?
- How it is related to other data?
- How it is used?
- How does it change over time?

Data scientists, analysts, and other business users are among the many stakeholders in your company who may greatly benefit from this visibility. Data lineage, on the other hand, contributes to metadata management by giving you important operational and historical context about your data. Your metadata will become more useful and actionable as a result. Standardizing and defining data sets according to business and tribal knowledge will be easier if you have a strong metadata management process in place that incorporates data lineage. Additionally, it may assist with data categorisation and effect analysis as data changes, promoting improved decision-making throughout your company(Ponnusamy & Gupta, 2023).

4.9 Chapter Conclusion

This chapter draws attention to the centrality of the advanced techniques in growing data engineering field to fit demands of a data driven landscape. Thus, it is possible to maintain data accuracy, consistency and availability by applying such techniques as efficient data modeling, high-quality ETL-process, and an integrated approach towards handling semi-structured and unstructured data. The combination of machine learning and data engineering shows that data pipelines can enable prediction models and real-time decisions. Additional features for managing work processes and metadata supporting operation and governance and improve scalability and compliance of data. These sophisticated methods prepare organisations for a challenging environment, for change, for competitiveness in a world where data volumes and variety are constantly increasing.

Multiple Choice Questions (MCQs)

1. What is the primary purpose of an ETL pipeline?

    a.   To visualize data

    b.   To automate data workflows

    c.   To archive data

    d.   To create machine learning models

2. Which database is best suited for handling unstructured data?

   a.   MySQL
   b.   MongoDB
   c.   PostgreSQL
   d.   Oracle

3. Workflow orchestration tools like Apache Airflow help in:

   a.   Building neural networks
   b.   Automating data pipelines
   c.   Real-time data transmission
   d.   Encrypting databases

4. What is the role of feature engineering in machine learning pipelines?

   a.   Cleaning the data pipeline
   b.   Automating data transformation
   c.   Extracting relevant features for models
   d.   Archiving unused data

5. Data lineage is important for:

   a.   Predicting customer behavior
   b.   Tracking the origin and flow of data
   c.   Visualizing data trends
   d.   Building dashboards

6. Which tool is used for containerization in data engineering?

   a.   Tableau
   b.   Docker
   c.   MongoDB
   d.   MySQL

7. Why is metadata management essential in data engineering?

    a. It speeds up batch processing

    b. It ensures data integrity and governance

    c. It replaces ETL pipelines

    d. It automates machine learning models

8. Which tool can be used for distributed data storage and real-time analytics?

    a. Excel

    b. Apache Spark

    c. Google Analytics

    d. SQL Server

9. Real-time prediction in data pipelines requires:

    a. Manual updates

    b. Batch processing

    c. Machine learning integration

    d. Data warehousing

10. What is a key advantage of using scalable data platforms?

    a. Reduced need for data security

    b. Handling growing data needs efficiently

    c. Eliminating the need for cloud storage

    d. Replacing real-time analytics

Answer

1	2	3	4	5	6	7	8	9	10
b	b	b	c	b	b	b	b	c	b

TRENDS AND INNOVATIONS IN DATA ENGINEERING

5.1 Chapter Overview

"Trends and Innovations in Data Engineering," provides an analysis of the developments that are currently leading to new changes in data engineering. The paper starts with the introduction of decentralized data organization approaches such as data mesh and data fabric, which are solutions to the problems of monolithic frameworks for big data. These approaches promote decentralisation of data assets but with centralised control and regulation. The chapter also overviews automation in data pipelines, demonstrating how automated actions, events, and management jobs enhance operational processes, enhance accuracy, and minimize the incidence of mistakes.

The use case of AI+ML within data pipelines is presented as a trend that will revolutionize the model operation, allowing real-time insights, scalability, and stronger model governance. Sustainability in data platforms is presented as a new trend, with an accent on using data to make climatically responsible decisions. In this chapter, the author explains how these trends working in synergy enable organizations to enhance the innovation, growth, and general resiliency of data environments to meet future demands.

5.2 Rise of Data Mesh and Data Fabrics

Data management may be difficult. Many businesses discovered that traditional, centralised data architectures need to be more adaptable in order to keep up with the rapid advancements in data, tools, and design. We will examine Data Mesh and Data Fabric in this part, along with their advantages and how they might assist businesses in overcoming the constraints caused by monolithic structures.

This straightforward illustration shows the conventional method of handling data and analytics:

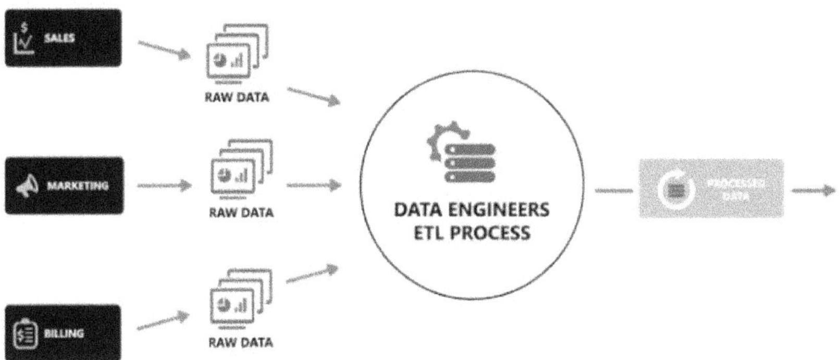

Figure 5.1: Monolithic Data System

Source: - *(HitachiSolutions, 2025)*

It's clear that universal management of security, quality, access, and other aspects is key when looking at this picture in the context of an IT-controlled data warehouse. But as usage rises, IT gets overloaded with demands for increasingly complicated data, which causes tension between the people who are providing and using the data. To put it simply, monolithic data platforms are not designed to grow with the constant demand for new insights. As businesses create and gather all of this data from both internal and external sources, they want a method for managing it that is both adaptable to changing needs and scalable to expand in a controlled manner.

In order to meet the increasing need for more effective and efficient data management, new design paradigms have developed. A data fabric is created by integrating this paradigm into a distributed architecture. With the ideas of treating decentralized data as a product and giving the organization's domains autonomy and ownership, the debate over how to handle dispersed data has developed into the data mesh.

Data Fabric

One entity serves as the hub of the data universe and orchestrates the movement of data between several systems, irrespective of format or location, in a data fabric architecture, which is a centralised method of dispersed data management. The goal is to offer a consistent and easy method for managing and accessing data across many devices, apps, and systems. Databases, data warehouses, data lakes, data catalogues, and data integration platforms are examples of current investments that are intended to be modernized.

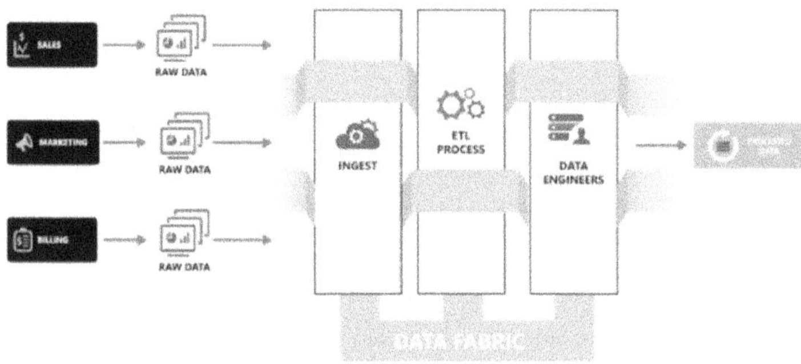

Figure 5.2: Data Fabric Architecture

Source: - *(HitachiSolutions, 2025)*

Consider it a platform ecosystem that serves as a self-service way to intelligently and securely coordinate various data sources. Data fabric design does not necessitate replacing and tearing down current systems. Existing apps and data storage contribute by giving the data fabric metadata. Once metadata linkages have been established, analytics are used to identify correlations between the data, not only the data you purposefully combined. The economics of data management have drastically altered, and insights are being generated on their own.

Data Mesh

A decentralized method for sharing, accessing, and managing analytical data within or across an organisation is the data mesh architecture. The goal of a data mesh is to match data products to organisational domains, such the departments that generate the data. According to this method, these domains are more likely to find value in their data. Furthermore, the domain that owns the domain is probably the subject matter expert, which allows it to handle the complexity and difficulties at their origin, including updates, backward compatibility, downtime, business semantics, and schema changes.

As seen below, sales-owned data practitioners assist each domain, like sales, in owning and managing its data and tools. In order to make the data discoverable, addressable, reliable, self-descriptive, and safe, the domain team uses product thinking. This guarantees that the individuals who know data the best can handle it effectively and satisfy the demands of the company.

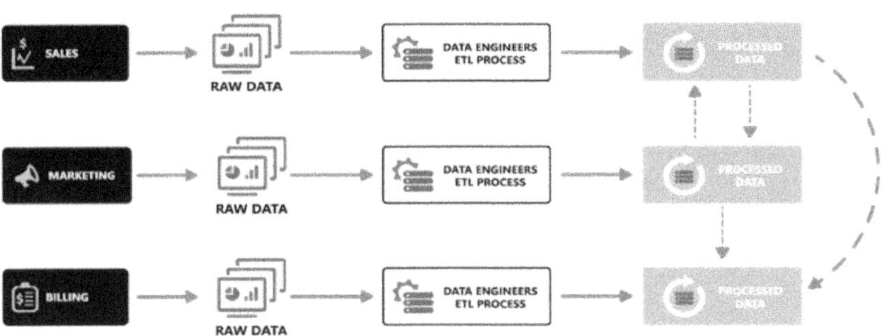

Figure 5.3: Data Mesh Architecture

Source: - *(HitachiSolutions, 2025)*

Through the autonomous ownership and management of data by several teams, data mesh seeks to provide a more scalable, adaptable, and secure data architecture. By giving data owners more freedom and flexibility and enabling more experimentation and innovation with data, data meshes address the drawbacks of centralised data warehouses. They also relieve data teams of the burden of meeting the demands of all data consumers through a single pipeline.(Akermi et al., 2023).

Combining Data Mesh and Data Fabric

Although some advocates of these new methods may be fervent about defining what "counts" as using data mesh or data fabric, we have found that in the few cases in which they are really used, they are rarely used in their "purest" form. Every organisation has a deeply rooted culture and set of limitations. A federal agency's decision about how best to use these strategies is influenced by a number of variables, including organisational structure and buy-in, data governance and metadata maturity, and particular operational requirements for data access, security, and agility. We can assist agencies in creating a hybrid strategy that provides a comprehensive solution to maximize data sharing while taking into account their unique organisational realities by maintaining the results they want to achieve from their data as the focal point of their work. In many situations, platforms that combine data fabric and data mesh might be crucial in laying the groundwork for adaptability and scalability to satisfy enterprise-level organisational demands.

Agencies may preserve centralised management and security requirements while enabling domain-specific teams to handle their data independently by utilising data mesh principles inside a data fabric architecture. By putting data producers in close proximity to the data they are most familiar with, this hybrid strategy enables agencies to leverage the agility and accountability that data mesh fosters. Concurrently, the centralised data fabric layer ensures uniformity and adherence to legal standards by offering standardised integration, governance, and security policies throughout the company.

This combination protects data security and integrity while enabling flexibility in data management and sharing to meet the demands of various departments or domains. Additionally, it promotes a culture of cooperation and creativity by making it possible for cross-functional teams to exchange ideas and work together efficiently, which improves the agency's overall capacity for data-driven decision-making. Federal agencies may adopt a balanced strategy to data sharing that maximizes data agility and governance by combining the greatest features of centralization and decentralization, which will eventually lead to increased operational efficiency and mission success(Hechler et al., 2023).

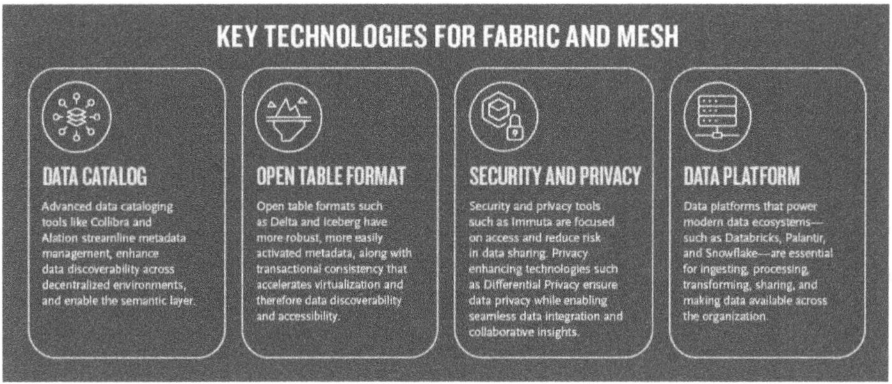

Figure 5.4: Key Technologies for Data Mesh and Data Fabric

Source: - *(Tucker, 2025)*

5.3 Automation in Data Engineering

Data engineering automation is the process of bringing automation to the otherwise manual steps involved in constructing the data pipelines used to extract, transform, load, analyze, and monitor data. Data pipeline creation is a multi-step process: It begins with data collection and extraction, then continues with transformations like cleansing, modeling, and masking as well as loading data for analysis, machine learning, and reporting.

Enterprises seek speed in decision-making, which is driven by data value. However, delays or long sync times can be detrimental to utilizing data for time-sensitive use cases, such as competitive analysis and customer-focused marketing. Furthermore, errors in construction or delays in processing can cause data to be outdated or obsolete by the time it arrives at its destination. Enterprises can expect better time to value by automating pipeline construction and management; they can also expect a decrease in expenses, errors, and delays throughout the process.

There are a number of common tasks that data engineers perform while building and maintaining data pipelines and systems. These tasks fall into four main categories: extraction, transformation, loading, and monitoring. Transformation itself encompasses three subcategories: data cleansing, data masking/sanitization, and data modeling. We focus on each task and present ideas and examples illustrating how automation and orchestration can help accomplish that particular task efficiently and reliably.

The table below summarizes these task groups and subgroups.

Data pipeline task type	How automation can help
Extraction	Data management tools periodically check a data source's status, automatically triggering a data extraction process using established connectors and returning standard output formats to make subsequent steps more efficient. Auto-generated connectors are a key innovation in this area because a lack of established connectors has been a common roadblock.
Loading	Automation can help load extracted data into target data stores for downstream consumers in a consistent, reliable way by developing data products that are capable of handling data with different formats and velocities.

Transformation	Cleansing	Automatic identification and correction (or removal) of incorrect or inconsistent data makes the data more reliable for downstream consumers.
	Masking and sanitization	Malicious code injected into data due to vulnerabilities in front-end systems or mistakenly entered sensitive data like credit card numbers can be automatically removed before causing security and privacy issues. Additionally, masking can help ensure that personal information is not shared with company personnel who shouldn't have visibility.
	Modeling	Modeling automation includes the automated creation of data products. Entire databases and tables, where key business objects are modeled, can be automatically modeled with consistent data types and naming conventions using script templates.
Monitoring		The increasing number and complexity of data pipelines leads to a requirement for constant and automated monitoring as well as immediate notification through various channels such as emails, Slack, etc..

This is not an exhaustive list of a data engineer's daily tasks. There are many other areas with tasks that engineers are sometimes expected to complete, such as addressing schema drift (the evolution of the data schema over time), but the same principles apply to them as well. The overall goal is to remove the need to do tasks manually in favour of building or buying solutions that perform them automatically. Automating these tasks reduces errors and enables organizations to scale their data engineering processes.

Having a flexible and extensible framework while building your data automation solution is particularly helpful in reducing cost and effort in the longer term.

Automation in common data engineering Tasks

Data extraction

How do you know your data is ready at the source to begin extraction? These are some commonly used methods:

- Manually refreshing a shared network or SFTP folder to see if the file you are looking for has arrived
- Running SQL queries and checking if they return anything for a given period
- Receiving a message, email, notification, or call from a colleague or system
- Receiving data from a continuous stream

Automatic triggers

In general, every time you must rely on manual work or notifications to start your pipeline—such as getting an email or message from someone telling you that data is ready to be extracted—there is an opportunity to automate. One way to automate the process is with automatic triggers, which are predefined conditions or events that initiate a pipeline when met. These triggers eliminate the need for manual intervention for notifications, allowing the pipeline to be automatically started when data is ready to be extracted.

For example, automatic triggers are helpful when you want to schedule the execution of your pipelines based on specific events, such as the arrival of new data, the completion of a previous activity, or a specific time or interval.

However, automatic triggers may not be suitable for scenarios that require complex dependencies or fine-grained control over the execution of tasks in the pipeline. Also, they can lead to issues such as resource contention or high costs if not adequately managed, mainly when multiple data pipelines are triggered simultaneously or when triggers initiate pipelines too frequently.

On the other hand, cutting-edge platforms such as Nexla have capabilities that extend beyond rule-based automation. The system is ready to identify and respond to anomalies. For instance, it is well-equipped to detect a sudden surge in requests or any activity that deviates from the norm. Not only does it monitor these irregularities, Nexla also leverages machine learning algorithms to understand, adapt, and make informed recommendations based on these unexpected occurrences. This dual capability of Nexla—implementing automatic triggers and detecting anomalies—ensures a comprehensive, intelligent, and efficient data engineering solution.

Control jobs

Another way to automate a data pipeline is by using control jobs: specialized tasks within a data pipeline or workflow management system that help manage and orchestrate the execution of other jobs or tasks. They control the pipeline's flow by ensuring that specific conditions are met before allowing other tasks to proceed. Control jobs can include monitoring data availability, checking the success or failure of previous tasks, or enforcing dependencies between tasks.

Control jobs are useful when you need to ensure that certain conditions are met before executing subsequent tasks. For example, you may need to wait for data to be available in a specific location or for an external API to return the expected results. By using control jobs, you can create more efficient and automated data pipelines. Implementing control jobs enhances the efficiency of data pipelines, taking automation a notch above what is conventionally possible.

While automation inherently minimizes the requirement for manual involvement, Nexla's features transcend this standard offering. Nexla fundamentally bolsters the reliability of the overall process, providing a superior level of stability and trustworthiness. In Nexla's framework, automation is not merely about replacing manual tasks but also fortifying the dependability of the data pipeline. This makes for a robust data management solution that promises consistency and reduces the risk of disruptions.

Note that control jobs can add complexity to the pipeline, requiring careful configuration and monitoring to ensure that they function as intended. Additionally, they can increase the overall execution time of the pipeline since tasks may need to wait for the control job to be complete or for specific conditions to be met.

Choosing between automatic triggers and control jobs

This choice depends on your data pipeline's specific scenario and requirements. For instance, imagine that you have an ecommerce website and need to process and analyze sales data daily to generate reports for the management team. In this situation, data is generated continuously as customers make purchases, and the analysis should be performed at the end of each day.

Using automatic triggers would be beneficial here because they can be set to initiate the data pipeline at a specific time (e.g., midnight) or when a particular condition is met (e.g., the daily sales data is ready). This guarantees that the analysis is completed on time, consistently, and without delays every day, removing the need for manual intervention.

Now imagine a sophisticated data pipeline in which data originates from several systems, each of which produces data at various times. In this scenario, tasks within the pipeline have complex dependencies on data availability, and the pipeline should only proceed when all required data sets are available.

In this case, control jobs can manage the dependencies between tasks. For example, control jobs can be configured to check for data availability at specific intervals and trigger the downstream functions only when the required data sets are ready. This strategy embodies both control and checks within the pipeline execution. It guarantees meticulous control over the sequence of operations, ensuring that tasks are launched only upon fulfilling their dependencies.

This dual strategy significantly lowers the possibility of mistakes and improves the overall dependability of the operation. It's not just about controlling the execution but also about validating the readiness and correctness of each step, creating a system that embodies both precision and accuracy.

It is essential to thoroughly weigh the advantages and disadvantages of each strategy in order to choose the best one for your specific use case. In some situations, combining control tasks with automated triggers may be necessary to get the automation and control you want in your data pipeline.

Data loading

A common issue when reading data from a source is having many different types of data. Some are traditional databases that return tables, some are streaming sources, and others are REST APIs that return data in JSON files or using other machine-readable formats. Occasionally, you may need to read data from log files that could be in a custom file format. In an ideal scenario, the output would be in a standard format that you can feed into downstream tasks; the same holds for target systems where you load the data.

One solution to this problem is using a polyglot approach that can handle data with different formats and velocities. Reading data from several sources, converting it into a common format, and then loading it into the target system are all made possible by a variety of data tools. However, this may be too time-consuming and resource-intensive for certain organisations.

There are solutions in the market that already solve this engineering problem, so you don't have to come up with a solution yourself. For example, an open-source framework like Singer's generalizes all data sources under the simple concepts of "taps" and "targets" to help streamline the process. However, its implementation requires time, resources, and technical know-how that you may not have available.

An ideal solution lets your data pipeline tool standardize data formats for you. For example, the Python Prefect library has Results as first-class objects, Azure Data Factory has Datasets as one of its main components, and Nexla has Nexsets as the building block for modern data architectures.

There is no need to invent and maintain a novel implementation when many tools already provide a solution.

Data Cleansing

The value of the output generated by any data system is directly correlated with the quality of the data that it ingests: "Garbage in, garbage out," as the saying goes. Data cleansing (or cleaning) ensures that any data that has any of the following issues in the table below is captured and immediately addressed before propagating further downstream.

When it comes to automating data validation and profiling, The market offers a wide range of technologies that can help data engineers with these duties. One such tool is Great Expectations, a Python framework that attempts automated profiling with a certain degree of success. However, it requires learning a lot of new concepts—like "data context," "expectation suites," and "checkpoints"—to perform a simple validation check.

If your data pipeline tool provides an easy-to-use feature for automated data validation, use it. If not, you can add simple inspection tasks to your pipelines and trigger them automatically when the upstream job completes. The method for adding simple inspection tasks and triggering them automatically will also vary based on the tool being used. Some tools may allow for automated data validation through scripting, while others may have a built-in feature for it.

Data masking and sanitization

Data engineers usually work with second-hand data. This is because the (initial) data collection happens at the application or device level, and the people who are involved are usually software engineers developing applications and database administrators managing relational databases. In an ideal world, data engineers will take care of sensitive data and malicious scripts (like Little Bobby Tables) before they creep into primary data storage systems. When data engineers find themselves in situations where data is not sanitized or needs to be shared within the company, they should develop automated scripts to deal with it before that data goes further downstream.

Before we delve into data sanitization categories and review some examples, it's worth noting that the terms "cleansing" and "sanitization" are often used interchangeably but have different connotations. Cleansing relates more to correcting errors in the data, whereas sanitization is primarily about hiding sensitive or malicious data matching specific patterns or characteristics (Mumuni & Mumuni, 2024).

5.4 Integration of Ai And Machine Learning in Data Pipelines

The process of creating safe, high-performing artificial intelligence (AI) applications is intricate and involves several steps. Development and operations teams require the workflow framework that the AI pipeline offers in order to operate in a methodical, regulated, and repeatable way. Pipelines enable the efficient and high-quality transfer of models from prototypes to production systems, from data preparation to model assessment. This section will examine the main steps of the AI pipeline and describe how machine learning operations (MLOps) automate development and deployment. We'll also go over the advantages of coordinating your ML activities with an AI pipeline.

<u>MLOps in AI pipelines</u>

MLOps applies DevOps concepts to the life cycle of machine learning projects. Data scientists, DevOps engineers, and IT may work together more easily thanks to a set of standardised procedures called MLOps. Building safe, scalable, and effective AI pipelines requires it. The operational and management procedures, version control, automated testing, CI/CD, training, and monitoring activities are all part of MLOps techniques, which concentrate on the particular tasks involved in creating and implementing models.

AI and ML pipeline stages

A organized workflow that regulates the creation, deployment, monitoring, and maintenance of models is necessary to create AI and ML systems that are suitable for production. This structure is provided by pipelines, which give a scalable, repeatable development process made up of several interrelated steps.

1. **Data collection:** The initial step in the pipelines is to collect the raw data needed to train the model's algorithms. Relational and NoSQL databases, data warehouses, APIs, file systems, hybrid cloud systems, and third-party providers are just a few of the many sources of data used at this stage. The sources and kinds of data gathered are determined by the project's commercial use case.

2. **Data cleaning and pre-processing:** Prior to being used for training models, raw data has to be cleaned and processed. Data must be analyzed, filtered, transformed, and encoded in this step of the machine learning process in order for the algorithm to understand it easily.

3. **Model training:** Data scientists fit the appropriate weights and biases to the method they have selected during model training, choosing the model and parameters that are most appropriate for their use case. Reducing the loss function, which takes into consideration the discrepancy between the target values and the model's anticipated outputs, is the main objective of model training. The team is guided by this performance metric as they strive to improve and enhance their model. A functional model that is prepared for testing and production deployment is the end product of model training.

4. **Testing and deployment:** A model must be evaluated to make sure its predictions are correct before it is made available to end users. In order to do this, a test set—a distinct subset of the data that was excluded during the training phase—is given to the model. The test set replicates the kind of real-world data that the model would probably come across after going into production. Data scientists may assess the model's ability to generalize and produce precise predictions on new data by assessing its performance on these previously untested cases. Potential problems with design, model selection, and programming are found and fixed using these findings. Once the necessary modifications are done, it is made available for usage.

5. **Model monitoring and updating:** Models need to be regularly updated and observed after deployment. To maintain an acceptable level of performance, problems including model degradation, data drift, and idea drift may need to be addressed on a regular basis.

Key Capabilities of an AI pipeline

Artificial intelligence is being used by data-driven organisations at a rapid pace and in an increasing variety of commercial use cases. The pipeline's capabilities become increasingly important as adoption picks up speed.

- **Efficiency and productivity:** An organized method for developing models is offered by artificial intelligence pipelines, which combine the efforts of several teams into a single, efficient process. Pipelines prioritise automation by eliminating the need for manual labour in processes like feature engineering and data preparation. Machine learning pipelines are compartmentalized, or divided into discrete, tiny parts, by design. Teams may swiftly test and enhance the pipeline design because to this architecture's encouragement of experimentation.

- **Reproducibility:** AI pipelines guarantee uniformity throughout projects with their highly automated, standardised procedures. A large number of the pipeline's components are easily reusable and adaptable to use in other projects.

- **Scalability and enhanced performance:** Pipelines for artificial intelligence can grow quickly to handle big datasets. Pipelines automatically distribute the computational burden among the number of processors needed for model training and inference using distributive processing. Many elements that enhance performance and efficiency are incorporated into pipelines, such as the ability to run multiple pipeline stages simultaneously and automatically optimize the distribution of computing and storage resources

- **Model deployment, monitoring and maintenance:** Pipelines are crucial for incorporating production-ready models into the applications, web services, or APIs that they will be used with. Model versioning allows developers to make necessary updates or roll back models. Developers may proactively identify permanent issues that will need to be fixed by retraining or updating software thanks to ML pipelines' stated protocols for evaluating post-production performance.

- **Iterative development and model evaluation:** Teams may experiment with different models and parameters as they look for the model or algorithm that best fits the issue requirements and data characteristics thanks to AI and ML pipelines, which promote an iterative development culture. Additionally, pipelines provide techniques for assessing model performance, which aid teams in determining the models' effectiveness both before and after deployment.

- **Model governance and security:** AI governance procedures guarantee that models are created and implemented in a morally righteous and open way that complies with applicable laws, company rules, and data security best practices. Checkpoints and validations are integrated into AI pipelines to guarantee that models adhere to these guidelines (Deekshith, 2019).

5.5 Decentralized Data Architectures

Conventional data architectures are frequently organized like mediaeval cities, with personnel, supplies, and resources housed in central areas that are simple to manage and protect. This isn't without justification, since your company can have consistent controls over who may access what data, when, and why by managing, storing, and retrieving data from a single repository.

The ongoing expansion of data and users, however, has exposed centralised systems' flaw: their inability to expand efficiently. Data-driven processes cannot develop and scale in inflexible, constrained environments, much as mediaeval aristocrats had to broaden their sphere of influence to take advantage of developing trade and agriculture. For teams that are expanding rapidly, this is particularly important since they want an architecture that can change with the business.

Instead of combining data storage and analysis skills into a single, centrally managed ecosystem, a decentralized data architecture divides these functions over several platforms and tools.

A decentralized network is "a network configuration where there are multiple authorities that serve as a centralised hub for a subsection of participants," according to the National Institute of Standards and Technology (NIST).

In the end, a decentralized architecture allows for dispersed data ownership and usage while preserving a certain amount of centralised control over data security and access. Different domains can be used to store and analyze data, and each one is maintained and controlled by an appropriate team or user inside your company.

A few essential elements of decentralized data architectures include:

- **Distributed Storage:** There is no longer a single, central site for data storage. Rather, the data is dispersed over several domains, each of which is maintained for a particular use case or purpose and overseen by the team or teams who interact with the data the most.

- **Federated Governance:** Data is not under the jurisdiction of a single centralised body since it is not centrally kept. While the central data security team establishes global regulations that all domains must follow, teams are responsible for developing and implementing local, domain-specific data access controls.
- **Interoperability:** Although people and data work in different realms, data shouldn't be totally isolated or unavailable. Decentralized architectures should enable cross-domain data exchange to promote collaboration, as users frequently need to collaborate across functional boundaries to advance business activities.

These elements may be combined by your team to construct a decentralized data architecture while maintaining some of the essential features of centralised systems, such as sufficient storage, data security, and accessibility.

Decentralized Data Architectures' Advantages

1. **Data Democratization:** Decentralization leads to data democratization, which means that more people in your company, regardless of function, may access data more easily and scalable. For companies that are expanding quickly, this is essential since more data users shouldn't put a strain on your data environment or the people who oversee it. Decentralization eliminates the requirement for new users to seek access from a single, centralised resource, allowing them to more rapidly obtain the domains and data they want.

2. **Reduced Management Burden:** This leads to still another advantage: central data staff will have less management work to do. Consolidated teams are frequently responsible for maintaining centralised architectures, which require them to examine and approve access requests from every user in an organisation. If this

team remains unchanged as the number of users increases, the workload associated with access management will increase rapidly. These centralised supervisors can exercise more effective control over the entire organisation when teams are in charge of their respective domains.

3. **Operability at Scale:** Last but not least—and maybe most significantly—decentralized systems are scalable. Compared to centralised ecosystems, which are ultimately constrained by their unity, this represents a substantial shift. Your company's data-driven operations may expand in tandem with your business requirements when domains are formed and managed as needed, becoming an interoperable component of the broader ecosystem.

Decentralized Data Architectures' Difficulties

1. **Lack of Consistency:** Teams may find it more challenging to guarantee the quality and consistency of data when it is dispersed across several platforms and domains. Inaccurate or flawed data from a domain team's poor-quality assurance efforts may affect not just their own data-driven choices but also those of any teams they are working with. Data consistency can also be harmed by a lack of interoperability since some domains could contain more recent resources than others.

2. **Siloed Data Resources:** Data silos may also become more likely as a result of decentralization. When teams operate completely independently of one another without exchanging or permitting access to the data stored in their own domains, silos like these arise. This can lead to conflict inside your company and increase the likelihood that stale or out-of-date data will produce erroneous analysis and insights. Teams must make sure that any technology change towards a decentralized data architecture is accompanied with organisational enablement and cultural buy-in in order to prevent silosing data and data consumers.

3. **Additional Security Risks:** Finally, there is some danger to the security and privacy of your data when decentralization is implemented. Decentralized architectures give different domain teams responsibility for this, but centralised systems might impose strict restrictions on all data. This necessitates a security-first corporate culture; otherwise, there is a danger of protective measures being slipped, leaving domains vulnerable to leaks, breaches, or abuse.

Role of AI Play in Enhancing Decentralized Data Architectures

Decentralized data architectures benefit greatly from artificial intelligence (AI), which automates quality control, integration, and data discovery procedures. In order to make better decisions, artificial intelligence (AI) systems can examine enormous volumes of data from dispersed nodes in order to spot trends, abnormalities, and insights. Moreover, AI may simplify data governance by automatically implementing standards and norms across domains, guaranteeing uniformity and adherence without the need for human supervision.

AI enhancements include:

- Automated data discovery and integration
- Advanced analytics for pattern and anomaly detection
- Streamlined data governance and policy enforcement

Impact of Decentralized Data Architectures on Data Security and Privacy

Data security and privacy may be improved and challenged by decentralized data systems. On the one hand, these architectures lessen the possibility of a single point of failure by spreading data over several nodes, which makes it more difficult for attackers to compromise the system as a whole. However, it might be challenging to provide uniform security protocols and privacy guidelines across all nodes. To reduce these risks and safeguard sensitive data, access restriction, effective data encryption, and frequent security audits are crucial.

Impacts on security and privacy include:

- Single points of failure are less likely.
- Maintaining consistent security measures might be challenging.
- Effective encryption and access control are essential.

Best Practices for Data Governance in a Decentralized Architecture

A decentralized architecture's effective data governance necessitates striking a balance between supervision and autonomy. Global standards may be established while giving domains the freedom to apply them whichever best suits their requirements by establishing a federated governance architecture. Compliance and uniformity across domains are guaranteed by routine audits and reviews. Furthermore, using technology to automate governance procedures may guarantee real-time policy enforcement and greatly lessen the workload associated with manual monitoring.

Among the best practices are:

- Adopting a federated governance model
- Conducting regular audits and reviews for compliance
- Utilizing technology for automated governance processes

Role of Decentralized Data Architectures in Driving Innovation in Data Management

Decentralized data architectures promote a more dynamic, robust, and scalable data ecosystem, which in turn spurs innovation in data management. Organisations may experiment with novel data strategies, quickly adjust to changes, and customize data solutions to meet domain-specific requirements by decentralizing data ownership and control. Teams are able to investigate and use cutting-edge data practices and technologies without being constrained by a centralised system, which fosters innovation.

Innovation drivers include:

- Encouraging domain-specific teams to try new things and be creative

- Scalability and adaptability to changing requirements
- customized data solutions for certain domain needs (Battiston, 2023).

5.6 Sustainability in Data Platforms

The sustainable data platform is a database system that contains a collection of reliable data for climate-neutral analysis and control. Since 2020, the platform for creating climate protection products has been developing in an agile and prototype manner. It seeks to make the causes and contributing elements of the climate catastrophe understandable and controllable.

In order to do this, it permits:

- Clear access to reliable data
- Understanding the ramifications of one's actions, both good and bad
- Accepting accountability for one's own behaviour
- Taking accountability for issues outside of
- Specific management and action

Goals and outcomes must be established in order to monitor climate protection progress and make targeted modifications as needed.

Participants create open-use tools on the open-source platform. In everyday life, the collaboratively created services, applications, or monitoring tools promote climate protection. Although quantifiable energy use and the corresponding CO_2 emissions will be the primary emphasis at first, other factors that contribute to global warming—such as consumerism, transportation, food, and infrastructure—will also be considered. For instance, people, governments, businesses, and building owners can use the platform's capabilities.

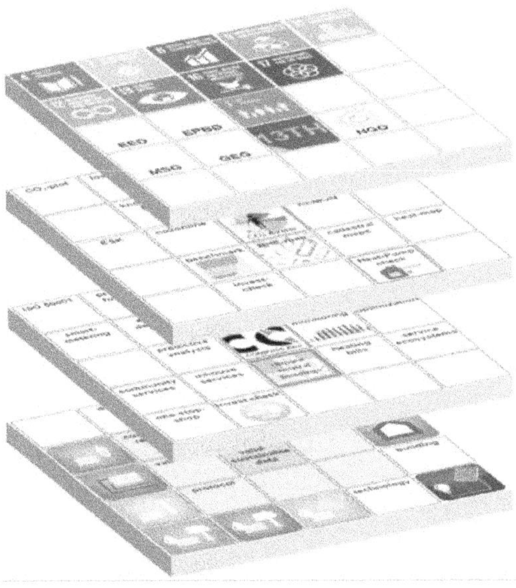

Figure 5.5: Structure of sdp with first modules and tools for bottom-up climate protection

Source: - *(SustainableDataPlatform, 2024)*

The tools are used to gather climate protection-related data from various participants, which is then processed for user input and anonymized for public feedback following validation. To provide outcomes in an atmosphere that is transparent and comparable, a scientific evaluation of the data that has been gathered is conducted.

The platform makes use of the growing digitization to automatically gather data, such as energy consumption ratings, renewable energy shares, traffic data, and cost projections. A pledge to utilize the data in line with the platform code is a prerequisite for participation.

Since 2020, a small group of dedicated stakeholders have worked together to establish the CO2COMPASS program and the sustainable data platform, both voluntarily and with financial assistance. Rapid prototyping and an agile approach to work are used to achieve a high development pace (Faggini et al., 2019).

Values of the Sustainable Data Platform

Values include decarbonization, civil society control over data and its processing structure, and transparent, democratic data management.

- **Valid data and transparent methodology for decarbonization**

To achieve carbon neutrality in a focused and controllable manner, the sustainable data platform creates and employs quantifiable, methodologically proven indicators and feedback mechanisms: For towns, corporations, zero-emission buildings, and individual and institutional carbon footprints.

Participants and municipalities will be given a solid foundation for political demands based on high-quality data. The platform encourages "learning from the best" and offers unbiased, comprehensive, and scientifically sound information on successes and failures using reliable, sustainable data.

- **Using digitalization for the common good**

The platform has an open data warehousing system framework. It is possible to receive services and feed in valid data from a large number of people. The system evaluates and stores data without being influenced by political or commercial interests or data abuse. SDP modules adhere to strict quality criteria for user experience and assistance.

The importance of the subject inspires commitment, and the modules provide the best possible rewards to sustain it. The "valid sustainable data" (programming interface for exchange) generated and used by the open platform can be removed at any moment by the data owners. Non-personal data, which is utilised anonymously in the context of the previously described decarbonization (open data), is not included in this.

- **Control of the platform by civil society**

To accomplish the aforementioned objectives, the platform assists actors and businesses in creating business models, such as institutions for empirical research.

By signing the platform code, the platform's players affirm the aforementioned principles and consent to independent civil society oversight by non-profit entities, including foundations with a charter that aligns with their objectives(Hellemans et al., 2022).

5.7 Chapter Conclusion

The progress outlined in this chapter demonstrates how data engineering has evolved toward more adaptive, elastic, and sustainable processes. Frequently applied tendencies such as decentralisation, automation, and AI integration allow organisations manage such huge data ambiguities, at the same time, enhancing effectiveness and inventiveness. This way businesses foster decentralized data management and gain operational scalability and flexibility at the same time as having strong governance and protection of their data. Disintermediation minimizes the participation of human activities, enhancing the dependability and expansiveness of data flows. In addition, AI and machine learning help to open the door for real-time decision-making as well as help data systems to adapt to the increasing need. Finally, examples of sustainability efforts in data platforms demonstrate the value of data engineering in handling the world's problems, including climate change. Combined, these innovations demonstrate the centrality of data engineering for building strong and adaptive organizations of the future.

Multiple Choice Questions (MCQs)

1. What is the primary focus of a data mesh architecture?

- a. Centralizing data storage
- b. Decentralizing data ownership
- c. Encrypting data pipelines
- d. Removing data governance practices

2. Which of the following is a key trend in data engineering?

a. Manual data ingestion

b. Decentralized data architectures

c. Limited automation

d. Monolithic data systems

3. How does automation in data pipelines improve workflows?

a. Increases manual intervention

b. Reduces errors and improves efficiency

c. Limits scalability

d. Decreases data processing speed

4. AI and machine learning integration in data pipelines enables:

a. Static batch processing

b. Real-time predictions and scalability

c. Manual feature extraction

d. Limited data insights

5. What is a key focus of sustainable data platforms?

a. Reducing security measures

b. Minimizing environmental impact

c. Increasing manual operations

d. Limiting data democratization

6. Which tool is commonly used for workflow orchestration?

a. Apache Kafka

b. Apache Airflow

c. MongoDB

d. Amazon S3

7. Data democratization means:

    a.   Restricting data access to IT teams
    b.   Making data accessible to all users
    c.   Eliminating data pipelines
    d.   Encrypting data for security

8. What is one challenge of decentralized data architectures?

    a.   Limited scalability
    b.   Complex data governance
    c.   Inability to handle real-time data
    d.   Slow data ingestion

9. What role does metadata management play in data engineering?

    a.   Visualizing data trends
    b.   Ensuring data security and tracking lineage
    c.   Replacing ETL processes

    4.   Automating dashboards

10. Which of the following is an example of a real-time analytics tool?

    a.   Tableau
    b.   Amazon Kinesis
    c.   Microsoft Excel
    d.   MongoDB

Answer

1	2	3	4	5	6	7	8	9	10
b	b	b	b	b	b	b	b	b	b

SECURITY AND COMPLIANCE IN DATA ENGINEERING

6.1 Chapter Overview

'Security & Compliance in Data Engineering' discusses the protection of data systems alongside following regulations in a technology world that is ever-growing. The chapter starts by warning about the risks of data breaches and unauthorized access to data and the need to encrypt data, control access, and store data securely. Other methods of access control such as RBAC, DAC, and MAC are discussed with emphasis on their role in controlling users' authority.

The chapter also takes time to explain how organizations should follow data privacy regulations including GDPR, CCPA, HIPAA, and PIPEDA, where organizations need to apply the concept of 'privacy by design' and ensure the privacy practices are disclosed. The disclosure of information, early identification, and evaluations as well as constant check-ups are described as promising concepts of organizational change in compliance with the law. Also, the chapter is dedicated to security incidents management and underlines that an organization should follow not only preventive but also reactive measures. Finally, procedures for a preventive security plan are discussed, which involves reminders of the penetrating test, incident response rehearsal, and a constant security awareness program to reduce the risk and improve stakeholders' confidence.

6.2 Importance of Data Security in Modern Platforms

The process of protecting digital information from corruption, theft, and unauthorized access during the course of its whole life cycle is known as data security. Hardware, software, storage devices, user devices, administrative and access controls, and the rules and processes of the organisation are all covered.

Data security makes use of technology and techniques that improve the visibility of an organization's data and its usage. By using techniques like data masking, encryption, and sensitive information redaction, these technologies help safeguard data. Additionally, the method assists organisations in complying with increasingly strict data protection rules and streamlining their auditing procedures.

An organisation may defend its data from cyberattacks by implementing a strong data security management and planning approach. Additionally, it helps companies reduce the possibility of insider threats and human mistake, which are still the main causes of data breaches.

There are several reasons why data security is crucial for businesses worldwide, across all sectors. Businesses are required by law to safeguard user and customer data against theft or loss and the improper use. Examples of state and industry regulations that outline an organization's legal obligations to protect data include the Payment Card Industry Data Security Standard (PCI DSS), the General Data Protection Regulation (GDPR) of the European Union, the California Consumer Privacy Act (CCPA), and the Health Insurance Portability and Accountability Act (HIPAA).

Preventing the reputational harm that comes with a data breach is another important function of data cybersecurity. A well-publicized breach or data theft may cause consumers to lose faith in a company and choose to do business with a rival. Significant financial damages are also possible, in addition to penalties, court costs, and damage restoration in the event that private information is lost (L. Sun et al., 2021).

Benefits of Data Security

1. **Keeps your information safe:** Sensitive information won't get up in the wrong hands if you adopt a data security-focused mentality and employ the appropriate technologies. Sensitive data might contain, among other things, identifying information, healthcare records, and client payment information. This information remains safe and secure with a data security approach tailored to your organization's unique requirements.

2. **Helps keep your reputation clean:** People entrust sensitive information to you when they do business with your organisation, and a data security policy allows you to give them the protection they require. Your prize? An excellent standing among customers, partners, and the business community at large.

3. **Gives you a competitive edge:** Data breaches are frequent in many businesses, thus being able to protect your data can help you stand out from the competition, which could be having trouble doing the same.

4. **Saves on support and development costs:** Including data security measures early in the development process will save you money later on when it comes to creating and implementing updates or resolving code issues.

It might be challenging to know where to start when it comes to your organization's data, even while data security is essential to keeping a contemporary data stack safe and efficient. By describing organisational and technological best practices for managing data security in contemporary situations, TDWI's new checklist report, Five Best Practices for Data Security in Modern Cloud Platforms, answers this need. These are the most widely used best practices:

1. Make Access Controls Compliant and Easy to Understand

Organisations must successfully manage access to this sensitive data in order to secure it. By doing this, you can make sure that only the appropriate individuals have access to and utilize sensitive data for the appropriate reasons at the appropriate times, and no more.

Additionally, companies are trying to make self-service data access possible so that users may easily get the appropriate access to their data. Simply put, access restrictions ought to make it easier to comply with rules without impeding setup or data usefulness. Writing clear, concise policies that legal and compliance stakeholders can understand can help ensure compliance. Laws, compliance rules, and standards, as well as moral data practices for the business, should all be followed by these policies.

Teams may keep an eye on whether their policies are being applied appropriately by implementing data access control methods. Additionally, data monitoring features keep track of who has access to sensitive data, when it was accessed, which data sources receive the most traffic, and which users utilize sensitive data the most frequently. This will assess the efficacy of the existing access restrictions and give internal and external auditors a better understanding of the compliance needs.

Businesses are attempting to democratize data use by making their data resources available to more people. Access to an organization's data sets is necessary for users in a data-driven culture, including data scientists, business analysts, and data engineers. Nobody wants to wait weeks or months to obtain the data they need, and teams don't want to be overloaded with requests for access that aren't being granted.

2. Consider Attribute-Based Access Controls Over Role-Based Access Controls

Each user is given a role or membership in an access group as part of role-based access control (RBAC), which establishes access. It is possible to divide this category according to department, location, or even title. A data scientist could have different rights than a data engineer, for instance. It is necessary to develop a new position with a unique set of rules if the data engineer need access to data that is outside of their purview. Because of a phenomenon known as "role explosion," which occurs when the number of additional roles that are needed increases rapidly, RBAC may become burdensome at this point.

The next significant development in access control is attribute-based access control (ABAC), which goes beyond what RBAC can do. Access is granted to people using ABAC according to a variety of contextual, data, and user characteristics. Objects like a file, the environment of the data, the activity being performed with the data, and the user's department or management level are a few examples. Instead of having to establish innumerable user-specific roles, data teams may provide access for different combinations of characteristics in this way.

As opposed to being applied statically like role-based rules, ABAC policies are dynamic and run at query time. ABAC rules may take longer to set up initially on the front end than RBAC since each policy must be created for each user. However, with time, these regulations become more flexible and granular, improving the overall performance of a data ecosystem. Regardless of technical proficiency, stakeholders may obtain the complete picture since ABAC roles are more logical and stated semantically.

3. Centralize Your Data Policy Management Capabilities

These days, businesses frequently have to handle enormous volumes of data from a variety of platforms, technologies, and contexts. Each cloud provider has its own platforms and apps, as well as its own security measures and access restrictions. Since data security and access control must be applied and maintained uniformly across different settings, the vast array of resources and platforms makes these tasks extremely challenging.

Organisations are seeking to consolidate and simplify their data policy management since so many policies are needed to safeguard and regulate access across various data ecosystems. Indeed, according to a recent research conducted by S&P Global's 451 Research, 52.5% of participants said their company used more policy-based data management procedures in an effort to obtain more reliable and consistent data.

Modern organisations are choosing to deploy single unified data security platforms (DSPs) since it is neither secure nor scalable to use several independent security solutions. Here, policies may be uniformly enforced across different cloud settings as they are developed, saved, and maintained on a single platform.(William McGeveran, 2019)several aspects of the law such as regulatory powers and penalties merit reconsideration. Some critics, however, have argued that the law makes the duty of data security inherently unclear—in the words of one legal brief, "an unknown (and unknowable.

6.3 Data Privacy Regulations

The GDPR, PCI-DSS, and HIPAA are three prominent examples of rules and regulations that have been implemented by governments and businesses to safeguard personal data. However, following in the GDPR's footsteps, rules like the CCPA, PIPEDA, POPI, and LGPD all pose significant challenges for businesses. The various data security and privacy laws will be examined in this article, along with how NetApp's Cloud Volumes ONTAP may enable compliance.

This is one of several comprehensive compliance management guidelines.

<u>GDPR Data Protection</u>

To strengthen and harmonise laws pertaining to the protection of personal data, the European Union passed the General Data Protection Regulation (GDPR). This thorough and unambiguous set of standards, which went into effect on May 25, 2018, recognizes that different "flavors" of personal data call for varying degrees of protection. The greatest degrees of protection are applied to sensitive data, including genetic, biometric, criminal, and health information. Companies that routinely gather and handle substantial amounts of personal data are required to register with government-appointed Data Protection Authorities, demonstrating that the volume of data matters as well.

All businesses that gather and handle personal data on EU citizens are subject to GDPR, regardless of their location. Non-EU businesses are responsible for all fines and penalties and must designate a GDPR representative.

Among the GDPR's primary obligations are:

- **Consent:** To acquire personal data, organisations need consent, and the degree of consent varies depending on the kind of data being gathered.
- **Data minimization:** The GDPR mandates that companies can only gather personal data that is directly tied to a clearly stated commercial aim, in response to years of applications collecting personal data without a clear reason. An organisation may not be in compliance if it collects personal information for one reason and then decides to use it for another (like consumer profiling).
- **Individual rights:** The GDPR's extremely explicit rights for data subjects—the people whose personal information is being collected—to know why and how their data is being processed are another important aspect of the law. They have the right to protest, to make corrections, and to be forgotten or deleted. Additionally, they are entitled to individual notification in the event that a breach of their personal data jeopardises their rights and freedoms.

The GDPR's "teeth," or extremely severe fines for non-compliance (up to €10 million or 2% of global annual sales, whichever is higher) and breaches (up to €20 million or 4% of global annual turnover, whichever is higher), are among its most distinctive features. The ability of data protection authorities to stop a business from gathering or using personal data while an investigation is underway into a potential non-compliance or breach is equally upsetting.(Truong et al., 2021)along with the blooming of Machine Learning (ML.

CCPA: The California Consumer Privacy Act

The rights of consumers to privacy are the main emphasis of the California Consumer Privacy Act (CCPA). The CCPA, which goes into effect on January 1, 2020, and is enforced by the Attorney General of the State of California, will govern personal information including cookies, IP addresses, internet activity, and biometric data. It will also regulate "household data" produced by IoT devices in the home, for instance.

Customers will have the right to know what personal information is gathered or sold, for what reason, and if prior transactions going back to January 1, 2019, are disclosed under the CCPA. They will be able to opt out of having their data collected or sold, see it, and request that it be deleted. Equal services at the same price will still be available to those who exercise these privacy rights. Additionally, customers will be able to sue businesses for privacy violations and data breaches.

The CCPA may apply to any organisation that may have the information of a California resident; failure to comply may result in fines of up to $7500 per infraction. Customers will now have the option to sue businesses for data breaches, seeking fines ranging from $100 to $750 per record.(Baik, 2020)

PIPEDA: Personal Information Protection and Electronic Documents

The Canadian federal privacy law for private-sector organisations is the Personal Information Protection and Electronic Documents Act (PIPEDA), which was granted royal assent on April 13, 2000. Its initial goal was to regulate companies that handle personal data in order to foster confidence in internet commerce. This rule is applicable to both foreign businesses that target Canadian consumers and any private company with a Canadian basis that gathers consumer data during business operations. Data gathered on a specific person, including name, age, ethnicity, medical history, opinions, remarks, and marital status, is subject to PIPEDA.

The 10 fair information principles that form the foundation of PIPEDA require companies to get consumers' consent before collecting any data. They must also restrict data acquisition to precise and well-defined goals and maintain transparent personal data rules. People are entitled to see their data and contest its correctness. Organisations that lose or steal data are likewise held liable by PIPEDA. Organisations covered by PIPEDA are required to notify the Privacy Commissioner of Canada and the persons impacted by a security breach of personal information as of November 1, 2018. Fines of up to CAD $100,000 might be imposed for failure to comply(Office of the Privacy Commissioner of Canada, 2019)use and disclosure of personal information, as well as for providing access to personal information. They give individuals control over how their personal information is handled in the private sector. In addition to these principles, PIPEDA (Personal Information Protection and Electronic Documents Act

LGPD: The Brazilian General Data Protection Act

The goal of the 2020 Brazilian General Data Protection Act (LGPD) is to replace and enhance current laws with a general data protection law that governs both the public and private sectors. By conforming to the global compliance requirements established by GDPR, it aims to boost Brazil's economy in addition to protecting personal data. Any organisation that gathers the personal data of Brazilians is subject to LGPD regulation, whether it is a tiny business or a big conglomerate. Protected data includes both personally identifiable information and anonymous information that can be used to determine a person's identity or for behavioral profiling

A Data Protection Officer (DPO) who communicates with Brazil's data protection body, the ANPD, and implements best practices must be appointed by entities covered by the LGPD. Businesses must also make sure that personal information is safe and report any potentially harmful data breaches to the ANPD. Customers will also be able to request that their data be changed, deleted, or transferred, as well as know why it is being gathered. Businesses who disregard these rules risk fines of up to $50 million Brazilian reals, or 2% of their entire yearly income in Brazil (De Lucca et al., 2023).

Australian Data Privacy Regulations

The Privacy Act of 1988, which combined federal, state, and territory legislation to govern the management of personal information, is the source of Australian data privacy rules. The private sector is subject to these rules, which are enforced by the Office of the Australian Information Commissioner (OAIC). By making sure that Australian companies with a revenue of more than AU$3 million adhere to specific compliance criteria, they want to secure customer data.

Australia has lately worked to update and enhance its current rules in the wake of the creation of the GDPR and other international legislation. According to the Notifiable Data Breaches system, certain Australian businesses are required to notify the OAIC of damaging data breaches as of February 2018. Furthermore, on November 26, 2017, the Australian government unveiled the Consumer Data Right (CDR), which gives customers in the banking, telecommunications, and energy industries the ability to access their data and decide with whom and for what reason it is shared. The Australian Competition and Consumer Commission (ACCC) enforces these rules, and breaking them can result in fines of up to $10 million(Von Dietze & Allgrove, 2014)substantially revised privacy laws came into effect in Australia following an extensive reform process. † The new laws introduce in many respects stricter data protection regulations alongside stronger enforcement powers, and if organizations have not already done so, now is the time for them to review and amend their data protection practices to ensure compliance with the new laws. † This article explains and analyses the major changes of relevance to private sector organizations (within and outside of Australia.

POPI: The Protection of Personal Information Act

Later this year, the Protection of Personal Information Act (POPI), which was signed into law in South Africa on November 19, 2013, is anticipated to take effect. Safeguarding personal data gathered in both public and private settings is its main goal. In order to guarantee responsible collection, storing, processing, and sharing of personal information, South African institutions are required to comply with a set of compliance criteria under POPI. While POPI is applicable to all South African businesses, it is particularly targeted at organisations like banks, insurance firms, and medical centres that manage enormous volumes of customer data. The legislation protects not just people but also any legally recognised organisation, such as businesses and communities.

Customers may access their data, request that it be deleted or modified, and manage who can access it under POPI. Businesses must gather data for legitimate and open purposes and keep it for as long as absolutely necessary. Additionally, they have to follow security compliance guidelines to make sure that neither they nor any third parties processing the data on their behalf hack or breach the data. Noncompliance with POPI regulations may result in penalties, jail time, and harm to one's image (Buys, 2017).

HIPAA Privacy and Security Rules

The primary goal of the Health Insurance Portability and Accountability Act of 1996 (HIPAA) was to promote the free exchange of health information in the United States in order to increase patient care outcomes and health care efficiency. National rules for protecting the privacy of personal health information were also imposed by these HIPAA compliance criteria. HIPAA's final Privacy Rule and final Security and Enforcement Rules have been mandatory since April 2003 and April 2005, respectively.

Health plans, health care providers, and health care clearinghouses are all considered "covered entities" for the purposes of the HIPAA rules and regulations if they communicate health information in writing, orally, or electronically. Additionally, it applies to business associates of covered businesses, which are people or organisations that are contracted to perform services but are not employed by the covered firm.

All "individually identifiable health information" that is transferred or maintained by a covered entity or its business associate, in any form or media—electronic, paper, or oral—is protected by the Privacy Rule, which is a little more expansive than the Security Rule. This protected health information (PHI) contains details about the person's physical or mental health, medications they have received, or the money they have been paid to receive those medications. Additionally, PHI contains basic identifying information including a patient's name, birthdate, Social Security number, and residential address. The purpose of the Privacy Rule is to promote health care research by allowing the use and exchange of de-identified health information without limitations.

The Security Rule only applies to e-PHI, or PHI that is stored or communicated electronically. According to the Security Rule, covered businesses must put in place the proper technological, administrative, and physical safeguards to:

- Make certain that every e-PHI they generate, receive, retain, or send is available, confidential, and intact.
- Determine and defend against security risks that are reasonably foreseeable.
- Prevent unapproved applications or disclosures that are reasonably foreseeable.
- Make sure their employees and business partners are complying.

To be considered HIPAA-compliant, cloud storage must meet all of these requirements. Compliance with HIPAA is supervised by the Office for Civil Rights (OCR). It has the authority to levy civil monetary penalties (CMP) for breaking the legislation, which can be as much as $1.5 million per event and range from $100 to $50,000 each impacted PHI record. In February of this year, for instance, OCR penalized Fresenius $3.5 million for five instances in which company did not adhere to HIPAA's risk management and risk analysis regulations.(Moore & Frye, 2019)and its regulations are complex. Many are familiar with the HIPAA aspects that address protection of the privacy and security of patients' medical records. There are new rules to HIPAA that address the implementation of electronic medical records. HIPAA provides rules for protected health information (PHI.

6.4 Secure Data Storage and Encryption Practices

Data is seen as an organization's most important asset as it defines what makes each business distinctive. It serves as the primary source of knowledge, information, and eventually the wisdom needed to make the right choices and take the right actions. It might be assisting in the treatment of a sickness, increasing a business's income, enhancing a building's efficiency, or taking accountability for reaching goals and raising performance. Additionally, data exchange, analysis, and storage are crucial services that each organisation needs in order to improve performance. However, businesses are under tremendous pressure to store the vast amounts of data locally due to the data's exponential evolution. The lack of resources has also made it more challenging to examine the data. Because of its numerous benefits, including on-demand service, scalability, dependability, flexibility, measurable services, disaster recovery, accessibility, and many more, the majority of enterprises have moved to the cloud for these services. A concept known as cloud computing makes it possible to have enormous amounts of memory and compute power at a relatively cheap cost. It gives cloud customers a great deal of ease by enabling them to access the desired services across several platforms at any time and from any location. Users may save money and increase productivity while managing projects and forming partnerships by moving their local data management system to cloud storage and utilising cloud-based services. As a result, more and more people and businesses are moving their various services to the cloud [8]. It is easy to predict that practically all enterprises will move to the cloud in the near future given the rapid advancement of cloud computing technology.

Despite the many benefits that cloud computing offers, there are a number of obstacles that might prevent its rapid expansion if they are not addressed. Think of a real-world example where a company allows its employees or departments to store and exchange data over the cloud. The organisation may be fully relieved of the responsibility of maintaining and keeping the data locally by taking advantage of the cloud. However, it also faces a number of security risks, which are the top worries for cloud users. First of all, sending data to cloud servers means that users have no control over the data, which makes them uncomfortable because the data may contain sensitive and important information. The second is that the cloud server was attacked, and data sharing is often implemented in an open and hostile environment. Under the worst circumstances, the cloud server itself may expose user data for illicit financial gain. In order to improve the performance of the business, the data must also be shared both inside and beyond the organization's premises with various pertinent stakeholders, such as consumers, employees, and business partners. Nonetheless, the recipient party may misuse this information and intentionally or unintentionally reveal it to an unapproved third party.

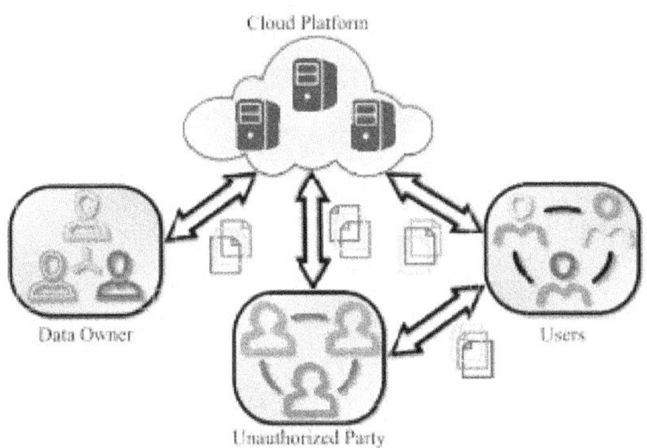

Figure 6.1: Block Diagram of Sharing Environment
Source: - (Gupta et al., 2022)

The aforementioned graphic depicts a sharing environment in which the data owners must transfer the organization's valuable data to the cloud platform because of the cloud's many advantages and the businesses' limited storage and processing capability. Additionally, for the sake of its usability, the cloud data is shared with several users based on their individual needs. However, once the recipient party has the data, they could leak it. Data may be stolen by an unauthorized person through illicit access, or it may be released by the people concerned. The organization's confidentiality may be seriously threatened by data loss or leaking. In addition to lowering the firm's rank and standing, it can destroy the enterprise's reputation and goodwill and lower the value of shareholders. Because data is a valuable asset for an organisation, it is imperative that this asset be kept safe. It becomes necessary to find solutions that can effectively safeguard data in a shared environment (Gupta et al., 2022).

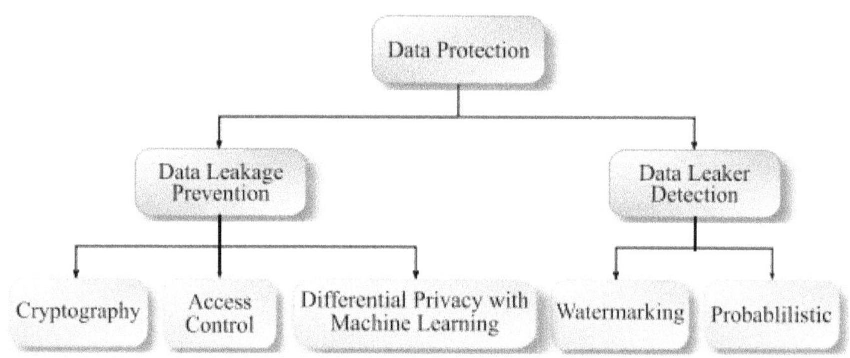

Figure 6.2: Data Protection Technique

Source: - *(Gupta et al., 2022)*

Secure data storage requires a risk analysis to build a network architecture adapted to your project and the type of users.

Viruses, Trojans, worms and ransomware are just some of the malicious codes to which your most sensitive data can be exposed, so applying measures such as access control or immutable storage is essential.

To familiarize yourself with these concepts so that you can enhance the security of your information, we are going to explain what is meant by secure data storage and the best practices that can be applied.

Secure data storage is the set of measures, configurations and criteria that ensure that stored resources are accessible only to authorized users and protects them from loss or theft.

In other words, it is all the resources that must be in place to ensure that the stored digital information is protected from any type of unauthorized access attempt.

It covers different areas of digital security, including network security controls and cyber threat analysis, physical security of equipment, communication protocols and user behavior.

The basic security features that a storage network must meet are as follows:

- It must be easily accessible to authorized persons and very complicated for any outsider.
- It must remain stable and secure, regardless of environmental conditions or specific contexts.
- It must have security mechanisms against all types of online threats of the order of ransomware, viruses, Trojans, etc.
- It must protect the integrity of the data and its availability to be modified, updated, etc. by the authorized person.

Also, beyond the data, secure data storage must be able to protect the privacy of the people who access it.

Secure cloud data storage covers the measures necessary to protect the integrity and security of information stored on remote servers accessed through an Internet connection.

It must meet the security requirements of any type of data storage with the added challenge of online connections.

To achieve secure data storage, the following best practices need to be considered:

Encryption of sensitive data

Data encryption is the basis for the protection of stored data.

For example, AES (Advanced Encryption Standard) encryption is a symmetric encryption that has been standardized and is often used to provide private and secure browsing in a Virtual Private Network (VPN).

Whatever type of encryption you use, it must be constantly updated and the keys must be stored in a secure location.

- **Access control**

Each device, software or infrastructure used for data storage must have an access control that discriminates, preferably, between access levels according to the use that will be made of the information.

Among the solutions that you can implement to avoid social engineering, multi-factor authentication is a very effective model that combines several methods of verification of the profile or system that tries to access the stored data.

- **Digital hygiene**

Digital hygiene comprises a series of practices and resources focused on protecting the digital identity of individuals that it is advisable to apply to data storage.

One of the measures is to remove all unused services from your storage system and the devices through which you access it to limit potential security breaches.

- **Immutable storage**

Immutable storage is a storage protocol that functions as a time capsule for documents, i.e., it keeps them intact for an indefinite or specified period of time.

While the information remains in this storage, it cannot be manipulated under any pretext to maintain data integrity.

However, it can be consulted. In other words, it is a write-once, read-multiple system.

Protection against the main detected threats

Before choosing an information storage system, it is advisable to carry out an analysis of the risks to which documents may be exposed by their nature, handling or exposure.

This makes it possible to create an adapted architecture based on the implementation of security layers, but also to select the necessary protection against possible threats such as viruses, worms or Trojans, malware or any other type of cyberthreat that can corrupt files.

- **Zero trust architecture**

Zero trust architecture is based on the assumption that every request for access to data, whether internal or external, must be validated and its legitimacy authenticated, regardless of the individual's background.

- **Informing users about security policies**

It is just as important to have secure data storage as it is for the people who are going to use it to know how to do so securely.

To this end, it is necessary to inform everyone who is going to work with the system or access it about data security policies through protocols appropriate to the nature of the information.

- **External tape storage**

External tape storage is a backup system based on the concept of data redundancy, a way of protecting information from loss or attack by keeping copies in different locations.

In that sense, this type of storage goes beyond local backup and recovery copies, since they are stored on tapes located in a different place from the main one, increasing their level of protection.

This makes it possible to recover information if it has been compromised by cyber-attacks such as ransomware.

- **Data Loss Prevention**

Many digital security experts advise using Data Loss Prevention (DLP) techniques to stop

information leakage, i.e., that any user can take internal information to external platforms or systems.

This type of system can be easily implemented through standardized templates.

- **Secure data transfer**

Secure data storage must also consider secure data transfer to guarantee information security using protocols like HTTPS (Hypertext Transfer Protocol Secure), FTPS (File Transfer Protocol Secure), SFTP (Secure Files Transfer Protocol) or MFT (Managed File Transfer).

6.5 Access Controls and Authentication Mechanisms

Developers that wish to secure their systems and safeguard sensitive data must comprehend access control mechanisms. The procedures and methods used to control and manage resource access inside a system are referred to as access control mechanisms. It entails figuring out who has access to what resources and when.

Developers may enforce security regulations, stop unwanted access, and lessen the chance of data breaches by putting access control mechanisms in place. Authentication, authorization, and different access control models like Mandatory Access Control (MAC), Discretionary Access Control (DAC), and Role-Based Access Control (RBAC) are examples of these techniques. Learn more about models of access control.

The methods and procedures used to regulate and oversee resource access inside a system are referred to as access control mechanisms. They are essential to maintaining the integrity and security of private information.

Who has access to what resources, under what circumstances, and with what degree of privilege is determined by access control mechanisms? They are intended to uphold security regulations, guard against data breaches, and stop unwanted access.

Depending on the particular demands and security requirements of a system, many kinds of access control mechanisms can be put into place. Typical varieties include:

- **Role-Based Access Control (RBAC):** According to the roles and responsibilities of people inside an organisation, this method grants access rights. It guarantees that users may only access the resources required for their job duties.
- **Discretionary Access Control (DAC):** Resource owners may manage who has access to their resources and set access permissions using DAC. Although it offers a great deal of freedom, it also has to be managed carefully.
- **Mandatory Access Control (MAC):** MAC uses pre-established rules and labels to enforce access controls. It is frequently utilised in settings like government or military systems where data secrecy is crucial.

Role-Based Access Control (RBAC)

One popular access control mechanism that distributes access rights according to predetermined roles and responsibilities is called role-based access control, or RBAC. It is appropriate for businesses of all sizes since it offers a scalable and adaptable solution to access management.

According to RBAC, particular positions inside an organisation are linked to access rights. Then, according to their work duties, users are allocated to one or more roles. Because administrators may now grant rights to roles rather than specific individuals, access control administration is made easier.

RBAC offers several benefits:

- **Simplified Administration:** The administrative strain of controlling each user's permissions is lessened with RBAC. Assigning or deleting roles makes it simple to give or remove access.
- **Granular Access Control:** Fine-grained control over access permissions is made possible by RBAC. To guarantee that users only have access to the resources required for their responsibilities, alternative roles can be developed to reflect varying degrees of rights.
- **Scalability:** Because RBAC is so scalable, it may be used by companies that see changes in their workforce or expansion. Adding or changing new roles won't impact already-existing access rights.
- **Security and Compliance:** RBAC ensures regulatory compliance and aids in the enforcement of security standards. RBAC lowers the risk of data breaches and unauthorized access by allocating access rights according to roles.

Roles must be defined, permissions must be linked to roles, and users must be assigned to roles in order to implement RBAC. To keep access rights in line with organisational requirements, it's critical to periodically evaluate and update job assignments.

Discretionary Access Control (DAC)

One access control mechanism that gives resource owners authority over access permits is discretionary access control (DAC). Owners of DAC have the ability to control who has access to their resources and to what extent.

The resource owner has the last say over access decisions when using DAC. The owner can provide access control to other users and has the power to allow or prohibit access to their resources.

Key features of DAC include:

- **Flexibility:** A significant degree of flexibility is offered by DAC as it enables resource owners to decide on access control using their own standards and judgement.
- **Ownership-based Access:** DAC is predicated on the idea that a resource's owner has the power to restrict access to it.
- **Access Control Lists (ACLs):** Access Control Lists (ACLs) are frequently used by DAC to define access rights for certain resources. A list that assigns certain access privileges to people or groups is called an ACL.

Although DAC provides flexibility, it also presents difficulties for centralised administration and ensuring uniform access control throughout a company. To guarantee that access rights are appropriately granted and withdrawn when needed, meticulous preparation and coordination are needed.

DAC is frequently utilised in settings like personal computers, file systems, and some collaborative systems where resource owners want control over their assets. It is not appropriate for highly regulated settings that demand centralised management and uniform access regulations.

Determining resource ownership, creating access control lists, and making sure resource owners have the resources and know-how to properly manage access rights are all part of DAC implementation.

Mandatory Access Control (MAC)

An access control mechanism called mandatory access control (MAC) uses labels and rules that have already been established to enforce access regulations. It is frequently utilised in high-security settings, such government or military systems, where data integrity and secrecy are crucial.

The classification and sensitivity of the resources as well as the users' security clearances are the basis for MAC access choices. It uses a hierarchical architecture in which access is either allowed or prohibited according to the resource's sensitivity and the user's security level.

Key features of MAC include:

- **Label-Based Access Control:** Labels are used by MAC to specify the security levels of resources and users. Classifications like Top Secret, Secret, and Unclassified are examples of labels.
- **Security Clearances:** Security clearances are given to users according to their access needs and degree of trust.
- **Strict Enforcement:** Users can only access resources with the proper security clearances because to MAC's stringent enforcement of access controls.

MAC offers a high degree of protection, but managing and implementing it can be difficult. The system's flexibility and agility may be restricted, and access regulations must be defined and enforced by a centralised authority.

MAC is appropriate for settings where data integrity and confidentiality are crucial, including systems that handle sensitive client data or classified information systems. By doing this, critical resources are protected from unauthorized users who lack the required security clearances.

MAC implementation entails creating security categories, allocating security clearances, and setting up the system to apply access controls in accordance with these classifications and clearances.

Authentication

The different authentication protocols that Security Verify Access supports may require users who wish to access your protected resources to supply credentials in order to authenticate. The Authentication Service is the part that is in charge of this functionality. In order to authenticate users, the Authentication Service includes a framework that you may use to enforce the usage of several available authentication protocols.

Modules known as authentication methods utilise a particular challenge or authentication technique, like a one-time password or user name and password, to authenticate the user. An authentication policy regulates the order the authentication processes are executed in. An XML document created using the authentication policy editor is known as an authentication policy. The sequence in which authentication procedures are executed is determined by the authentication policy.

The Authentication Service oversees the implementation of the event's necessary authentication policy during an authentication event. The authentication policy workflow includes each authentication technique. The Authentication Service authentication policy initiates this process. The Authentication Service creates a user credential after the user has successfully authenticated to each of the authentication methods specified by the authentication policy. At the point of contact, this user credential establishes an authorised session for the user.

By setting up the authentication policy, the administrator may decide what data is contained in the credential. An administrator can define the properties to be included in the generated credential using the credential editor provided by the authentication policy editor.

The credential that was produced includes:

- **Authentication Types** and **authentication Mechanism Types** attributes to indicate the authentication policies.
- Procedures for authentication that are finished during the authenticated session.

The administrator can use **authentication Types** and **authentication Mechanism Types** attributes to author an access control policy to require:

- The user must authenticate once in accordance with the policy using an authentication policy or method.
- Every time the user accesses a protected resource, they must authenticate.

Two execution flows are supported by the Authentication Service:

- Implementation of an authentication event following an assessment of access control.
- Request authentication from the user because of either:
- A point of contact access control list (ACL).
- A protected object policy (POP) evaluation.

Authentication mechanisms

The following authentication methods are available through Security Verify Access:

One-time password authentication mechanisms

A password that is created for an authentication event and is only good for one usage is called a one-time password. The following functions are offered by Security Verify Access's one-time password authentication feature:

- One-time password creation and verification that works with a variety of implementations.
- Delivering passwords once by email and implementing short messaging services (SMS).
- One-time password creation and validation using time-based, counter-based, and RSA techniques that don't need a delivery method.

Username and Password mechanism

Users supply their password and user name.

- **HTTP Redirect mechanism:** Incorporate a custom authentication method into an authentication policy's workflow by using this approach. Users supply the credentials that the customized authentication system needs.
- **Consent to device registration mechanism:** Users provide permission for their device to be registered.

- **FIDO Universal 2nd Factor mechanism:** FIDO Universal 2nd Factor tokens that have been registered are used by users for authentication.
- **FIDO2/WebAuthn mechanism:** Registered FIDO2/WebAuthn authenticators are used by users to authenticate.
- **FIDO2/WebAuthn registration mechanism:** Users can sign up for WebAuthn and FIDO2 authenticators.
- **Decision JavaScript mechanism:** During branching policies, this method is utilised to execute JavaScript rules(Singh & Singh, 2023).

6.6 Ensuring Compliance in Data Engineering Workflows

1. Understanding Regulatory Requirements

Compliance in data engineering begins with a thorough understanding of regulatory frameworks relevant to data processing and storage. These might include industry-specific rules, the California Consumer Privacy Act (CCPA), the General Data Protection Regulation (GDPR) for users in Europe, or the Health Insurance Portability and Accountability Act (HIPAA) for data pertaining to healthcare. Data engineers must interpret these regulations to design systems that align with legal requirements, focusing on aspects such as user consent, data retention policies, and data sovereignty. Periodic audits of these interpretations help in maintaining alignment with evolving regulations.

2. Implementing Data Privacy by Design

A compliance-focused workflow integrates privacy at every stage of the data lifecycle, commonly referred to as "privacy by design." Data engineers must anonymize sensitive data, use pseudonymization techniques, and implement secure encryption methods during storage and transit. By embedding these principles into the architecture, the risks of data breaches are minimized, and regulatory compliance is enhanced. For instance, masking personally identifiable information (PII) ensures that even in the case of unauthorized access, the exposure of sensitive data is prevented.

3. Role-Based Access Control (RBAC)

Ensuring compliance requires robust access management practices to prevent unauthorized data usage. Role-Based Access Control (RBAC) makes sure that workers only access the data they need to do their jobs by granting access to data according to designated roles within an organisation. Data engineers need to implement access hierarchies and maintain activity logs to monitor data usage. Regular reviews of access permissions reduce the likelihood of internal data misuse or accidental violations of compliance protocols.

4. Data Quality and Integrity Standards

Compliance is also dependent on maintaining high data quality and integrity standards. Accurate, consistent, and validated data ensures compliance with regulatory reporting requirements. Data engineers must establish validation checks, de-duplication protocols, and error detection mechanisms within workflows. A comprehensive data lineage system allows engineers to trace and verify data transformations, ensuring transparency and accountability in data handling processes.

5. Compliance Monitoring and Automated Auditing

One essential element of contemporary data processing is the integration of automated systems for compliance monitoring. These technologies have the ability to continually examine workflows and datasets for any violations, identifying abnormalities instantly. Automation reduces human error, increases efficiency, and provides an audit trail for regulatory review. Open-source and commercial tools like Apache Ranger, Collibra, and BigID are often integrated into workflows for compliance monitoring.

6. Data Governance Frameworks

In data engineering, compliance is supported by a strong data governance system. The rules, procedures, and guidelines for handling organisational data are outlined in this framework. It guarantees adherence to compliance standards by all parties involved, including data engineers and end users. A culture of responsibility is also promoted by a governance architecture, which guarantees that data engineering teams coordinate their activities with corporate objectives and legal mandates.

7. Secure Data Storage and Transmission

Secure transmission and storage must be given top priority in data engineering processes to guard against legal violations and online threats. Encrypting data while it's in transit and at rest guarantees that private information cannot be accessed by unauthorized parties. Engineers must also use secure protocols like HTTPS and SFTP for data exchange. Regularly updating software and implementing intrusion detection systems adds another layer of security.

8. Incident Response and Recovery

Data breaches can happen even with precautions, therefore having a clear incident response strategy is essential. For breaches to be handled quickly and legally, data engineers must work in tandem with the IT and legal departments. To increase future resilience, incident response procedures should incorporate post-event evaluations, data recovery techniques, and communication methods for impacted users and regulatory agencies.

9. Employee Training and Awareness

Compliance is a shared responsibility that extends beyond technical measures. Organizations must conduct regular training sessions to educate employees about compliance requirements, data handling protocols, and the consequences of non-compliance. By fostering awareness, data engineering teams can minimize the risks associated with human error or ignorance.

10. Staying Updated with Evolving Regulations

The regulatory landscape is constantly changing, requiring data engineers to stay informed about new laws and updates to existing frameworks. This can involve subscribing to industry newsletters, attending compliance workshops, and consulting legal experts. Proactive adjustments to workflows based on emerging regulations help avoid costly compliance lapses.

By integrating these strategies into their workflows, data engineers can build systems that are not only efficient and robust but also aligned with the highest standards of compliance. This alignment not only protects organizations from legal penalties but also fosters trust among users and stakeholders (Martin & Kung, 2018)for Privacy by Design to be viable, engineers must be effectively involved and endowed with methodological and technological tools closer to their mindset, and which integrate within software and systems engineering methods and tools, realizing in fact the definition of Privacy Engineering. This position will be applied in the soon-to-start PDP4E project, where privacy will be introduced into existent general-purpose software engineering tools and methods, dealing with (risk management, requirements engineering, model-driven design, and software/systems assurance.

6.7 Handling Security Incidents and Breaches

The process of identifying, evaluating, controlling, and reacting to security hazards inside an organisation is known as security incident management. It seeks to restore business continuity while reducing the harm caused by security incidents such as hacking, data breaches, cyberattacks, and system outages. The methodical procedure that IT teams employ from the moment a breach occurs until regular operations are restored is known as security event management. Four steps are typically involved: identify, analyze, mitigate, and restore.

The security incident management system creates the replies in real time, and the events might be intentional efforts to compromise the company's IT infrastructure or just unintentional exposures to security threats.

Security events have the potential to demoralize stakeholders, impair the organization's information assets, interfere with operations, harm its brand, and have legal repercussions. One risk management tactic to protect the company from these outcomes and handle security breaches early on is security incident management.

The three ways to approach security incident management are as follows:

- ***Reactive incident response***

Under this approach, security incidents are addressed as and when they arise. It begins with determining the causation of the event, containing its impact, and initiating steps for mitigation. A post-incident review is then carried out for system improvements.

Conducting a forensic analysis after a data breach to identify the root cause is an example of a reactive incident response.

- ***Proactive incident response***

Proactive incident response is an approach where preventive security measures are taken before the occurrence of an event to tighten the security of the IT infrastructure. The aim is to nip issues in the bud and stop them from becoming more serious.

Periodic risk assessments and vulnerability scans are examples of proactive incident response approaches.

The Hybrid Approach: Best of Both Worlds

While proactive measures like data encryption, avoiding unauthorized access, etc are necessary, it is not possible to ward off security incidents completely. The chances of security accidents still prevail and that is why, there also must be a reactive incident response approach.

The best solution is a hybrid approach where the organization takes measures to prevent security breaches in the first place while having a plan for incident response in case of event occurrence.

The process of handling incidents varies depending on the business context, type of incident, and policies, among many other factors. Generally speaking, the initial steps start with the investigation of the system or applications that display anomalous behavior. Common indications of a suspicious intrusion include a slowed-down system, inability to access files, and frozen screens.

The security administrator will investigate these malfunctions to determine if the system is breached, in which case, they further analyze the type of attack, the depth of damage, and the right steps to contain it.

Finally, the team implements the right steps to mitigate the incident and restore the system to ensure business continuity.

The goal of having a cybersecurity incident management framework is to reduce the response time and control the damage. So, it is important to have a well-structured approach when laying down the framework.

Sprint connects with your cloud stack to monitor the control checks and categorizes it into failing, passing, critical, and due in a single dashboard. This way, you get a centralized view of your assets to prioritize critical failures that may escalate into an incident and manage them on priority.

Here are the 6 steps to ensure security incident management:

1. Outlining the scope

The first step is to lay down policies and procedures for security incident management. The scope of events covered, types of data, and the required team should be defined. The organizations should also spell out what infrastructure will be needed, what kind of training must be provided and what kind of updated situational awareness is vital for incident handling.

The scope of your incident response strategy should be updated if you add any new regulatory compliance programs or add new incident response technologies to your system.

2. Communication channels

Depending upon the nature and size of the organization and the complexity of operations, there can be different departments that directly or indirectly contribute to security management. A proper incident response team with analysts and IT professionals must be employed with clearly specified functions. Then, there can be a legal and compliance team for ensuring security compliance, a PR team for communications etc.

Having communication channels to keep the trust of stakeholders and customers intact is also very crucial. In case of an event occurrence, internal communication for explaining the action plan and external communication for maintaining transparency about compromises should be smartly handled.

3. Incident detection plan

An incident identification procedure helps pinpoint early warning signs of malicious activity. Any unusual activity is highlighted through alerts and notifications. Red flags are then analyze for entry source and severity of impact on business functions. The analysis determines the prioritization of incident handling.

Your incident response process and critical decisions should be approved by the senior management. It should also address the potential impact of future attacks in the business environment.

4. Risk containment, eradication, and recovery

This is the stage where all the major work is done. Risk containment involves taking immediate steps to prevent the spread of the threat, risk eradication aims at treating the underlying cause, and risk recovery incline towards business continuity restoration.

So, if a network or system is infected, isolating it from other systems or shutting down services will be risk containment measures that will help control the spread. Forensic analysis will then be carried out to determine the root cause of the incident, followed by recovery measures like restoration of files from backups.

5. Post-incident retrospective

A retrospective meeting should be conducted post recovery to discuss the wins and gaps. How did the issue arise, what was the immediate response of the team, what was missing in the collaboration efforts or infrastructure wise and what went well, everything should be discussed and documented?

6. Continuous monitoring and testing

As mentioned above, both reactive and proactive measures ensure security incident management. So, there should be periodic assessments and testing of security systems, threat scans, activity log checking, security audits etc. The purpose is to stay vigilant and put reliable security protocols in place for solid cyber resilience (Chockalingam & Maathuis, 2022).

6.8 Best Practices for Proactive Security

There could be cybersecurity teams out there that are made up of elite experts with the greatest equipment, and for them, money is essentially nothing. Regretfully, the majority of us do not inhabit that reality. Rather, we work in a world where resources are few, trade-offs are ongoing, and skill and experience levels vary.

This reality frequently shows itself as a security staff that is unable to keep up, trapped in a reactive posture, following down a barrage of alarms and failing to react promptly enough to events. This is a surefire formula for employee stress and potentially harmful data breaches over an extended period of time.

However, by being more proactive, one may foster organisational learning and get a better comprehension of hazards and the most effective ways to avert them. So how can one manage risk in a sustainable way and confront issues before they arise?

To keep a proactive security posture, follow these five recommended practices.

- **Develop a Strategy that Balances Active Interventions with Prevention**

A solid strategic framework is the first step towards effective risk and vulnerability management. Clarifying the type and worth of the assets you're safeguarding, figuring out the most popular attack techniques, and comprehending the types of flaws your system has previously had are all crucial.

Equipped with this understanding, you are able to execute the subsequent action: Implementing preventative security procedures to find vulnerabilities in your current security setups.

- **Adopt the Mindset of an Attacker**

This is essential to have a better grasp of your security posture's actual advantages and disadvantages. By taking this method, you may examine how your defenses are functioning in the real world rather of focussing on how they should function in a hypothetical, ideal setting. You may recognize and get ready for the most likely assault routes and strategies your enemies will use against you by gaining this viewpoint.

- **Identify and Assign Defined Roles**

A team is more likely to fail if its roles, duties, and power structures are unclear. Making sure that no lines are crossed and that each team member is aware of exactly when to intervene and what to do is crucial if you want to recognize dangers and address them effectively. You will be at the forefront of proactive threat protection if you have clearly defined roles and duties.

- **Avoid Box Ticking**

Excessively reactive security postures are characterised by box ticking activities. All too frequently, security experts feel they are protected by the technologies they install for particular problems. The issue is that those instruments are frequently either misused or inadequate for the job. Go further and make sure that the appropriate individuals are ready to utilize your tools, that they all function effectively together, and that they fit into the larger strategic framework of your defenses in order to prevent this.

- **Test Your Incident Response**

Include frequent incident response exercises in addition to vulnerability assessments and pen testing (automated or human). Combining all of these components guarantees that you are taking a comprehensive strategy and that you will be ready for any attacks that arise in the real world, which is an important factor in proactive cyber defense.

Organisations nowadays require a better understanding of their own resources and the dangers that threaten them. The best method to do this crucial duty is to create a smart proactive security posture and enhance it using cyber-attack simulation and other state-of-the-art security solutions.

Benefits of Proactive Security

A proactive strategy to security is preferable over a reactive one because of the tremendous harm that a data breach can do to income and reputation. While proactive security can guarantee data protection, assist compliance, and prevent exploits before they occur, reactive security has a purpose and can also be used. Proactive security also saves the company money by preventing assaults and harming its reputation.

Proactive security has various advantages for organisations. The decrease of risk to revenue, brand reputation, and productivity is the primary advantage, but there are a number of others as well.

- Emergency containment and cleanup are no longer a continual distraction for your developers and operational staff. Continuous crises reduce output and create a stressful work environment for employees. Attacks are prevented via proactive security, allowing dangers to be examined rather than contained.

- Prevent data loss and breaches. The organisation does not have to handle incident response or deal with the fallout from a breach since proactive security prevents breaches and prevents attackers from accessing data.

- New dangers can be found and investigated. As dangers are discovered, attackers take the opportunity to modify their code and discover new weaknesses. Your company is ahead of the curve with proactive security, enabling it to take the required precautions to shield data from recently identified threats.

- Determine weaknesses before they are discovered by adversaries. Penetration testing is a component of proactive security, which means your company will identify flaws before attackers can take use of them. Sometimes even the greatest systems have ignored vulnerabilities or incorrect setups, but proactive security detects them early.

- Continue to comply. Data monitoring is necessary for compliance, and proactive data security strategies assist businesses avoid paying steep fines for data breaches.

- Cut down on incident response and investigation expenses. You have fewer probes into potential breaches because proactive security prevents assaults. Although incident response and investigations are costly and time-consuming, investing in these processes may save money.

- Boost consumer trust and loyalty. A company that has many data breaches loses the trust of its clientele. You may gain more customers' trust and draw in new ones by being proactive in protecting data so that your company doesn't make news for data breaches.

Proactive security research demonstrates that when an organisation adopts a proactive strategy, it improves all aspects of attack detection, protection, identification, response, and recovery. Proactive security reduces the chance of a data breach, saving money and brand reputation, according to every statistic (Huth & Nielson, 2019).

6.9 Chapter Conclusion

This chapter emphasizes that in the age of digitalization, data security, and compliance are proactive guideposts of data engineering strategies. In other words, measures of security can significantly reduce the likelihood of threats, safeguard data and credibility of an organization. The features related to access controls, data storage, and adherence to best practices and regulatory standards help align with international standards and factor in user confidence in the service. Furthermore, the presence of a proactive security approach enhances the organization's capacity to manage new threats, in addition to developing efficient incident response frameworks enhances an organization's ability to manage new threats. In this context the dynamics of regulation dependency and changes, keeping abreast of these changes and being ready to adapt in order to maintain the overall business integrity and prove compliance with the principles of ethical use of data. Altogether such approaches represent the common proven base for proper and future-proof data engineering protection.

Multiple Choice Questions (MCQs)

1. What is the primary goal of data encryption?

    a.   To automate data pipelines

    b.   To protect data from unauthorized access

    c.   To improve data visualization

    d.   To increase data velocity

2. Which access control method assigns permissions based on roles?

    a.   Discretionary Access Control (DAC)

    b.   Mandatory Access Control (MAC)

    c.   Role-Based Access Control (RBAC)

    d.   Data Transformation Control (DTC)

3. GDPR and CCPA are examples of:

    a.   Data visualization tools

    b.   Data security protocols

    c.   Global data privacy regulations

    d.   Data encryption algorithms

4. What does "privacy by design" refer to?

    a.   Building systems without security measures

    b.   Integrating privacy measures from the start of system development

    c.   Adding privacy features after deployment

    d.   Limiting data access to IT teams

5. Which strategy helps detect unauthorized access to data systems?

    a.   Batch processing

    b.   Continuous monitoring

    c.   Data replication

    d.   Data visualization

6. What is a hybrid approach to data security incidents?

    a.   Using a single security protocol

    b.   Combining preventive and reactive measures

    c.   Eliminating all security checks

    d.   Relying solely on encryption

7. What is the purpose of regular audits in data compliance?

    a.   To limit data availability

    b.   To ensure regulatory compliance

    c.   To increase data storage

    d.   To automate ETL pipelines

8. Which of the following is a key element of secure data transmission?

   a. Using public Wi-Fi networks
   b. Encrypting data during transit
   c. Disabling firewalls
   d. Removing user authentication

9. What is the role of incident response drills in data security?

   a. To automate data workflows
   b. To test the effectiveness of security protocols
   c. To limit data access
   d. To increase data ingestion speed

10. What does proactive security involve?

   a. Reacting after a breach
   b. Anticipating and preventing potential threats
   c. Removing access controls
   d. Avoiding data encryption

Answer

1	2	3	4	5	6	7	8	9	10
b	c	c	b	b	b	b	b	b	b

PRACTICAL INSIGHTS AND INDUSTRY APPLICATIONS

7.1 Chapter Overview

This Chapter expands on how the concepts of scalable data platforms are applied in practice across different industries and how data engineering enables business transformation at scale. The chapter discusses how companies use patterns of scalability like distributed computing, cloud services, and containers to process big data in real-time. Some of the examples include Netflix that built an application, based on microservices; Amazon that developed the scalable database systems; and LinkedIn, that optimized the real-time data processing models.

It also presents important use cases of real-time analytics in various industries such as financial, healthcare, manufacturing, logistics and retail. Use cases include fraud detection, predictive maintenance, dynamic pricing, recommendation system, precision medicine and public health. The chapter focuses on the issue of addressing obstacles or difficulties for scale-out data engineering system design and implementation of data ingestion, integration, storage, and processing. A wide range of best practices such as Cross-functional collaboration, Automation of workflows and DevOps are mentioned to handle data operations. Moreover, it offers information on new and exciting jobs within data engineering, including what data architects, big data engineers, and machine learning engineers do.

7.2 Real-World Implementation of Scalable Data Platforms

Scalable data platforms are a revolutionary approach for organizations that need to process enormous amounts of data in real-time. In the real world, these platforms are crucial for business organizations in terms of providing insights that will help them to make improvements in operations and indeed customer experience. For instance, there is the use of cloud solutions, for example AWS and GCP and Azure that afford scalability, meaning organizations can add or subtract resources based on the amount of work required at a given time. This elasticity is especially useful when the amount of traffic and transactions can grow during such activities as sales or holidays in the case of an e-commerce company.

In order to process vast amounts of data over several nodes, large-scale data platforms can include advanced technologies such as distributed computing frameworks like Apache Hadoop and Apache Spark. This makes it possible to have maximum availability and also to be fault tolerant which is very important for such sectors as finance and health where reliability of data is paramount. Also, the new DBMS like NoSQL (MongoDB, Cassandra, etc.) facilitate flexibility in managing unstructured data types like videos, feeds, IoT sensors data, etc., which are used more and more in manufacturing, media, etc.

Containerization technologies such as Docker and Kubernetes are also used by organizations to deploy a microservices architectural model. This approach enables it to scale individual services without any hindrance and balancing the resources that make up the whole system. For instance, Netflix, which is an online streaming company uses it to make the content recommendations based on the massive terabytes data in a real-time environment.

The microservices-based architecture used by Netflix is well-known for dividing big apps into more manageable, standalone operations. This enables Netflix to grow many aspects of its platform, including content recommendation algorithms, video streaming, and user authentication, as needed. It is possible to launch and grow each microservice separately and horizontally.

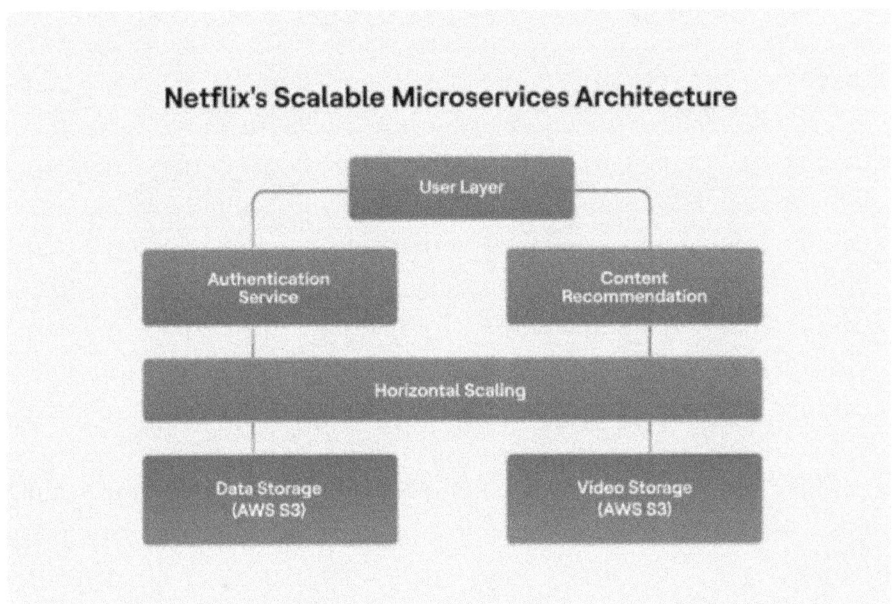

Figure 7.1: Scalable Data Architecture of Netflix

Source: - *(ACTIAN, 2025)*

Amazon's extensive e-commerce infrastructure, which processes millions of transactions every day, is supported by its scalable database design. Amazon effectively manages its inventory, client orders, and transportation by using horizontal scalability, data segmentation, and sharding.

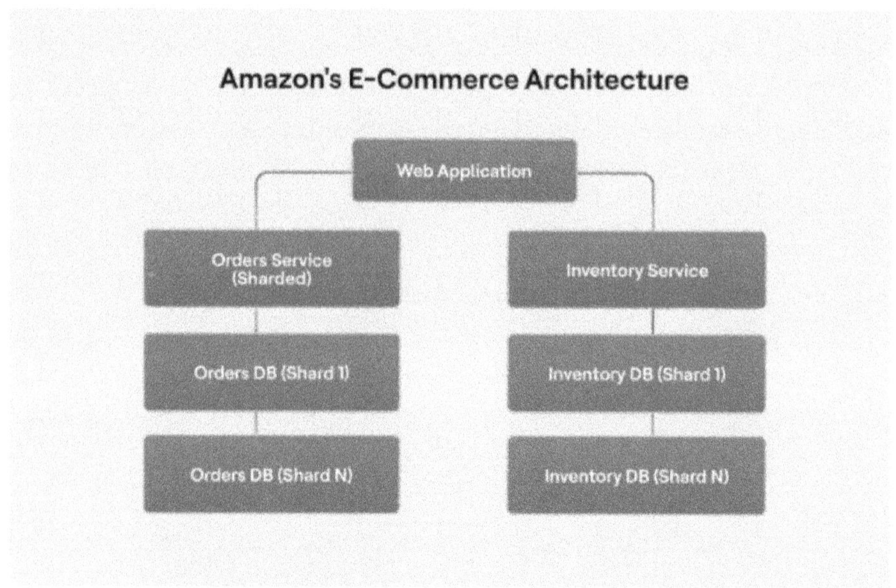

Figure 7.2: Scalable Data Architecture of Amazon

Source: - *(ACTIAN, 2025)*

This design divides work across several smaller databases, enabling Amazon to manage the massive and fluctuating workloads of its e-commerce business.

The architecture of LinkedIn is built to manage massive data input and processing in real time, guaranteeing that users get changes (such job suggestions and news feed items) immediately. Hadoop is used by LinkedIn for distributed data storage, and Apache Kafka is used for real-time data streaming.

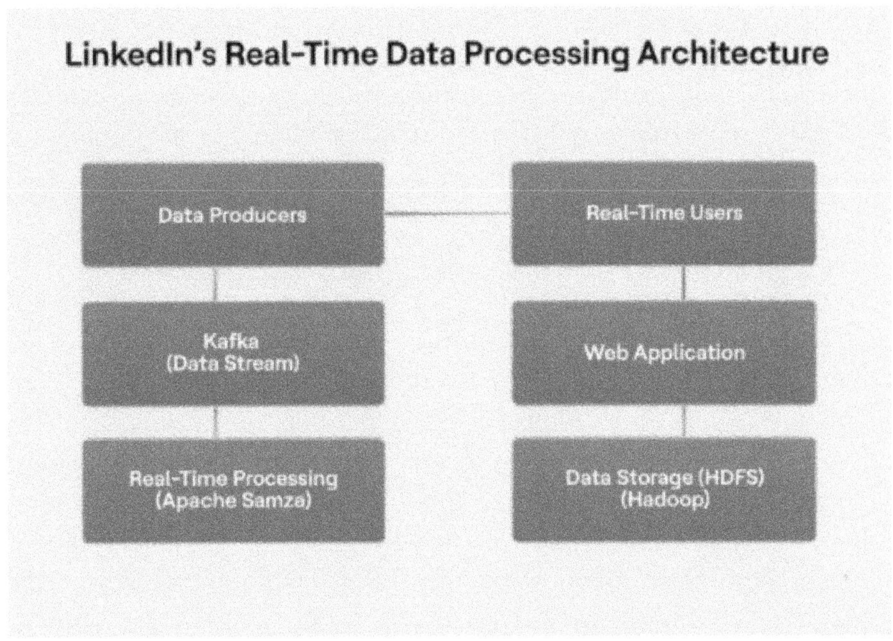

Figure 7.3: Scalable Data Architecture of LinkedIn

Source: - *(ACTIAN, 2025)*

The use of these models has been expanded through their combination with scalable data platforms such as Artificial Intelligence (AI) and Machine Learning (ML). For instance, organizations such as Uber use such sites to anticipate high demand and hence plan the most effective routes to prevent inconveniences hence enhancing performance. The same goes for retail behemoths such as Walmart to run predictive analytics for things like pricing and stock management.

In addition, the scalable data platforms are equipped with effective data security necessities to meet the global requirements such as the GDPR and the HIPAA. By employing encryption, role-based access control, and audit trails in these platforms, value-creating information is protected, which is reassuring to stakeholders.

7.3 Key Use Cases in Real-Time Analytics Across Industries

Real-time analytics examples in financial services

1. **Regulatory compliance:** "The federal and state governments have strict regulations governing the mortgage business. Customers must thus be informed of specific disclosures. And failing to do so can lead to severe fines in addition to undermining customer confidence. With the correct technology, mortgage lenders may better understand when and how to make the proper (or inaccurate) disclosures in all customer contacts. Before audits, mortgage lenders can identify those encounters and address the problem, especially when mandatory disclosures aren't being made. Employees who consistently fail to provide the necessary statements might also benefit from improved coaching and training. Reducing risk and adhering to compliance rules improves your bottom line, which is always crucial but becomes increasingly more crucial as the business develops.

2. **Credit scoring:** For instance, Equifax created actionable explanations that are customized for each customer by integrating machine learning modelling (neural networks) into an explainable artificial intelligence credit scoring algorithm. There are other agencies experimenting with machine learning techniques besides Equifax. Experian added machine learning features to its analytics tools to produce more in-depth, real-time insights. In order to identify high-risk identity behaviours and produce more precise, comprehensible scorecards for credit applications, TransUnion and FICO also integrated machine learning. Even for "credit invisible" customers without recently updated credit files, the more recent Vantage Score employs machine learning to evaluate risks and provide ratings. Platforms for generating machine learning models are being developed by other agencies, like Credit info.

3. **Financial trading:** "Measuring risk as a last step is no longer sufficient in quickly evolving financial markets. In March 2008, traders having exposures to Bear Stearns learnt the hard way that trading choices may drastically change exposures in a moment. Capital markets are increasingly shifting to intra-day value at risk calculations in order to evaluate market portfolio risks and implement real-time corrective actions.(Antova et al., 2019).

<u>Real-time analytics examples in manufacturing & logistics</u>

1. **Responsive processes in transport logistics:** "The sector is increasingly in need of improved visibility along transportation routes. To guarantee higher productivity, shipping managers are altering the way they run the lines. From the beginning to the end, real-time freight analytics are essential. Managers may swiftly address any gaps and inefficiencies by using the constant flow of real-time data. Simultaneously, the data systems offer more frequent and thorough updates, which leads to automatic updating and notification of all parties involved. By implementing an automated and responsive approach that utilises real-time freight information, the supply chain may be completely impervious to disruptions.

2. **Predictive maintenance:** Businesses in the industrial sector may utilize predictive maintenance analytics as a tool to better predict equipment breakdown and save needless downtime. In order to assess the functioning of equipment both now and in the future, predictive maintenance analytics combines real-time equipment data with data analytics and machine learning. Businesses may better comprehend equipment status and see potential warning signs of an impending breakdown by keeping an eye on equipment conditions in real-time. Businesses may then use this data to create a data-driven maintenance plan and decrease equipment downtime to boost output.

3. **Production planning:** Overall equipment effectiveness (OEE), one of the most potent measurement instruments in an industrial setting, is driven by real-time analytics. By exposing all production losses, automation experts are able to make unbiased business decisions that enhance plant asset performance, capacity, and utilisation. It may assist businesses in achieving their goals, including shorter turnaround times, higher throughput and quality, more predictable supply chains, and lower expenses. When properly assessed and implemented, OEE enables managers to make prompt, accurate, and objective judgements.

4. **Fleet management and driver safety:** "A range of data types, including telematics data, data from different cloud or edge devices, GPS, vehicle cameras, traffic cameras, and driver monitoring programs, are the foundation of fleet management data analytics. The information gathered can help identify the most frequent reasons for mishaps and provide guidance on how to prevent them. For instance, if reckless driving is the primary cause of collisions, the issue can be resolved with an appropriate training program. Alternatively, fleet managers can utilize predictive maintenance to anticipate incidents before they happen and stop them from happening if the vehicles themselves are the source of the accidents (Parida, 2024).

Real-time analytics examples in retail & eCommerce

1. **Omnichannel experiences:** Many merchants lack the time and resources necessary to integrate and consolidate their data across all channels of client engagement.

   Nevertheless, the main reasons why customers are concerned are fragmented data and a disjointed customer experience. Eighty-nine percent of consumers say they become irritated when they have to describe the problem again to an agent or in chat. Use technologies such as conversation intelligence driven by AI to map out and comprehend the complete client experience in order to reduce these risks. A crucial first step in bridging intelligence silos and equipping agents with knowledge when clients switch across channels is customer journey mapping.

2. **Dynamic pricing:** "Offering competitive prices is crucial to maximising revenue without driving customers away from your retail business.".

"To manage offers, use dynamic pricing. It's a way to adjust pricing to reflect the market. To find the optimal prices for a store, current dynamic pricing uses machine learning, artificial intelligence, and historical data. Dynamic pricing is constantly used by ride-sharing companies such as Uber and Lyft. Days with poor weather or during rush hour have an impact on service expenses in order to generate more revenue (Braik et al., 2016).

Real-time analytics examples in marketing, social media & digital technology

1. **Real-time agent guidance:** Artificial intelligence (AI)-powered conversation analytics and conversation intelligence software may help you enhance the frontline agent experience by offering real-time monitoring and recommendations that agents can use to maximize every contact. To help you better understand your consumers and the factors that influence their behaviour, conversation analytics tools such as Call Miner examine each encounter across a variety of media. You can foster a culture of self-improvement by using Call Miner to provide employees real-time feedback and advice on the best course of action to help them transform bad contacts into positive customer experiences.

2. **Artwork and image selection:** Marketing, according to some, is more art than science. Both are married when it comes to the visual images that Netflix utilises to draw in users. Based on the age and general tastes of each user, Artwork Visual Analysis (AVA), "a collection of tools and algorithms designed to surface high

quality imagery from videos," may anticipate which merchandise will still appeal to them the most.

3. **Virtual reality:** Facebook claims that "positional tracking that [is] precise, accurate, and available in real-time" is essential to the operation of their Oculus Quest virtual reality (VR) and augmented reality (AR) headsets. Additionally, according to the business, this positional tracking system has to be small and energy-efficient enough to be used as a stand-alone headset (Santos et al., 2023).

Real-time analytics examples in healthcare

1. **Benefit verifications and prior authorizations:** "CVS Health, a leader in hyper automation, has long optimised its support services by utilising RPA, AI, and other business process automation tools. For instance, CVS Health was able to automate their prescription intake, benefits administration, and revenue cycle management procedures by using AI, RPA, machine learning, data analytics, and natural language processing (NLP).

2. **Medication adherence:** In order to keep patients on specialty treatments, which are complicated and frequently cause adverse effects or unsuccessful therapy, regular monitoring and individualized involvement are necessary. The patient may quit taking the medicine if the adverse effects are severe or if they no longer believe the drug is helping them. Saving payors money requires proper monitoring to determine if this is occurring and the implementation of suitable actions.

3. **Faster, more accurate breast cancer screening and diagnosis:** "A fresh collection of real-world data from iCAD, a company that develops 3D mammograms, demonstrated that their AI-powered screening tools were able to reduce the number of false positives and enhance the detection rates of breast cancer. ... In contrast to the results of a group of radiologists, researchers discovered that the average number of tumours diagnosed per 1,000 patients screened increased from 3.8 to 6.2 when the AI system was installed.

Additionally, the percentage of outcomes that were incorrectly interpreted decreased from 9.6% to 7.3% (Wills, 2014).

7.4 Overcoming Challenges in Large-Scale Data Engineering Projects

Data Ingestion

When implementing a data engineering approach, one of the initial engineering challenges often encountered is related to data ingestion. The main problem is that data comes from a variety of sources and has different forms and patterns. As a result, data must be transformed before it can be further processed and examined.

Furthermore, real-time data ingestion poses potential challenges in data engineering as it demands high-speed processing. Establishing effective and scalable data input systems that can manage massive data volumes and process them in real-time is crucial to addressing this.

Another significant data problem in this field is guaranteeing data integrity and quality assurance. Analysis and insights may be faulty as a result of inaccurate or inconsistent data. Thus, it becomes essential to create data validation and cleansing procedures in order to identify and address problems with data quality during the intake phase.

Data Integration

When embarking on a data engineering project, a significant engineering challenge often arises in data integration, particularly concerning the relationship between data sources and software solutions. Integrating data from several systems and effectively connecting disparate information sources are the main goals of any data engineering project. This task can be particularly daunting when dealing with outdated legacy systems that lack the inherent capabilities to connect with modern software.

To address this data challenge effectively, it is advisable to prioritize the modernization of legacy software before delving deeply into data engineering initiatives. By undertaking this modernization process at the data engineering project's outset, potential integration complexities can be minimized in the future.

Additionally, apart from dealing with disparate systems, integrating data often involves managing different formats, structures, and semantics. Consequently, to guarantee compatibility and coherence throughout the linked dataset, data transformation, mapping, and schema alignment become crucial.

Data Storage

There are two main engineering issues in the field of data storage. The first concerns adapting to the increasing amounts of data. Data storage solutions need to be able to scale easily in order to handle this. As data requirements grow, data engineers may take advantage of technologies like cloud-based storage services and distributed file systems, which provide simple scalability while preserving performance levels and avoiding exorbitant expenses.

Organizing and retrieving data is the focus of the second data challenge. Handling enormous volumes of data dispersed over several systems can make organisational tasks more difficult and impede quick and effective data retrieval. Optimizing data access patterns and decreasing retrieval times requires the use of practical techniques such data indexing, splitting, and careful data structure design.

Additionally, data engineers should investigate data encoding and compression strategies to improve storage space economy without sacrificing data accessibility or integrity.

Data Processing

In the realm of data management, the continuous generation of digital information poses a significant challenge for businesses. The sheer volume of data from sources like mobile apps and IoT devices can be overwhelming, necessitating efficient processing techniques to handle this influx effectively.

Data engineers frequently utilize distributed computing frameworks such as Apache Hadoop or Apache Spark to address the challenge of processing massive amounts of data. These frameworks improve the speed and scalability of data processing processes by enabling parallel processing over a cluster of devices.

Moreover, data integrity issues, such as incomplete or erroneous data, can compromise the accuracy of analytical outcomes. Inconsistencies in data across systems, especially without real-time updates, can lead to inaccuracies that hinder business insights.

Developing a data governance plan in conjunction with a data management strategy is essential to tackling these data engineering difficulties. By delegating accountability for all data-related tasks and putting regulations in place to protect data integrity, this strategy guarantees the dependability and caliber of digital information inside the company.

Data Quality and Governance

Maintaining data dependability and quality is crucial in the field of data engineering in order to prevent problems. Effectively identifying and resolving problems with data quality requires the ongoing application of data validation and cleansing techniques, such as outlier identification, data imputation, and validation rules.

Moreover, regulatory compliance poses another significant challenge, especially for businesses in sectors like finance and healthcare. Regulations like HIPAA, PCI DSS, and GDPR can impact operations, requiring strict adherence to data-related laws.

To navigate these challenges in data engineering, a combination of strategies is recommended. Staying informed about evolving regulations, potentially seeking legal counsel, and Working with data engineering experts who are experienced in creating compliant platforms can assist guarantee compliance with the most recent regulations and industry best practices for data management.

Data Pipeline Orchestration

Data pipeline orchestration presents a complex process involving multiple stages and interdependencies, posing a significant challenge in coordinating and managing data processing tasks across diverse systems or components.

The complexity is increased by the presence of data dependencies between various processing steps or activities, where the results of one activity flow into another. It might be difficult to manage these dependencies and guarantee that the necessary data inputs are available on time.

Moreover, when dealing with data pipelines, issues like network failures, hardware malfunctions, or processing errors may arise.

Data engineers employ fault-tolerant architectures, rely on strong orchestration frameworks, and make scaling plans in order to overcome these obstacles. Incorporating monitoring and troubleshooting tools is also essential. These strategies facilitate efficient and reliable data processing, ensuring a seamless flow of data through the pipelines (Cao, 2010) demonstrating, and pushing the use of specific algorithms and models. The process of data mining stops at pattern identification. Consequently, a widely seen fact is that 1.

7.5 Industry-Specific Applications

a. Healthcare

As technology develops further, data engineering's applications in healthcare become more and more numerous. It is possible to optimize the whole operational chain and provide patients with individualized healthcare with the correct patient information.

Every data requirement in an organisation may be supported by data engineering. Long-term sustainability and viability require an effective data engineering approach.

Predictive Analysis for Disease Prevention and Precautionary Steps

According to statistics, even in different people, disease-causing microorganisms have comparable life cycles. Doctors can choose the best course of action based on their prior diagnosis by gathering and preserving data on a big population.

Additionally, machine learning and deep learning are being used more often in the healthcare industry. For instance, it is more common to use scientific studies on how neural networks increase the prediction of specific diseases.

The TOP5 criteria can be used by researchers to determine the five most likely health issues. The degree of organisation and storage of the data will determine how accurate this study is. Following the listing of the most likely illnesses, medical professionals can develop a plan for diagnosis and treatment. This results in a quicker resolution and patient release by preventing misdiagnosis.

The likelihood of a speedier diagnosis can also be greatly increased by having access to the family medical history. Therefore, hospitals must use data engineering in 2023. This technology is already being used by a few hospitals in the US. To assist guarantee a seamless rollout, the reputable John Hopkins Hospital is collaborating with GE and the Cleveland Clinic's Microsoft partnership.

Patient Care and Record Management with Personalized Treatment

The future of medicine lies in customized treatment regimens. According to researchers, it will be standard practice in healthcare globally by 2030. Patients have the option to provide access to their information to healthcare organisations. Electronic Health Records, or EHRs, are widely used in the world's healthcare system. Data engineering in healthcare makes the information conveniently processable with patient permission. Providence St. Joseph Hospital, Children's Mercy Hospital in Kansas City, and NHS UK are a few early adopters of data-engineered personalized care.

Operational Efficiency and Resource Optimization

In order to lower the possibility of human mistake, hospitals all around the world are moving towards using data to inform business choices. You may utilize analytics and data engineering to forecast and replenish hospital inventory before shortages happen if you have accurate historical data about how your hospital operates. Data engineering is already being used by Chinese hospitals to control their supply chains.

Additionally, hospitals may employ data engineering to better analyze patterns and be ready for future health-related situations. These data analytics algorithms are used by Assistance Publique-Hôpitaux de Paris (AP-HP) to forecast the number of patients who will visit them each month for both emergency and outpatient care.

Vaccine Research and Clinical Trial Management

Researchers and healthcare professionals learnt from the pandemic how crucial big data analytics is to vaccine development and clinical trials. Many institutions share information widely, which enables us to get the intended outcomes more quickly. Improved vaccination research was made possible by data engineering. Pfizer utilises Snowflake Data Cloud to assure accurate diagnosis and enhance patient outcomes more quickly through predictive analysis. They leveraged Snow grid's capabilities to provide their several teams, spread throughout the globe, a single source of truth (SSOT) so they could make the best data-driven choices possible. This made sure that everyone was aware of the most recent developments in vaccines and that all clinical trial data was readily available.

Public Health Management and Outbreak Prevention

In many nations, public health committees are in place to offer recommendations for ensuring the highest possible level of health for their population. These committees are in charge of keeping track of those who have been diagnosed with various illnesses. Public health systems can find isolated cases before they become a health issue by using data engineering models. "An ounce of prevention is worth a pound of cure," as the proverb goes. The Data Integration Partnership for Australia (DIPA) is used by the Australian government's healthcare department to detect adverse occurrences. Big data analytics is also being used by Norway to monitor changes in national health (Wang & Alexander, 2020).

b. Finance

Risk Management

- **Use case:** Data engineering facilitates the collection and synchronization of many types of financial data into a single data repository. Credit ratings, transaction logs, financial data, and other economic measures are only a few examples of these many formats. A unified approach to thorough risk modelling and evaluation guarantees consistency in the analysis and comparison of data by leveraging data engineering best practices.
- **Benefit: Data** Engineering provides a precise and accurate risk assessment by streamlining data integration and normalization. The use of data engineering can also make it possible to predict market movements and anticipate possible risks.
- **Example Use Case:** By using data engineering principles to integrate market data and transaction histories, JP Morgan Chase enables institutions to conduct optimal risk management analysis using effective dashboards generated from real-time data.

Fraud Detection and Prevention

- **Use case:** Significant real-time data is necessary for an efficient machine learning system for fraud detection to be accurate and produce fewer false positives. Algorithms will be able to identify irregularities and stop fraudulent transactions before they occur by examining spending habits and behaviour. By creating a single source of truth, data engineering helps reduce the likelihood of fraud by training fraud detection algorithms to identify abnormalities and outliers. A fraud detection algorithm requires both historical and real-time pattern monitoring in order to identify anomalous activity in advance.

- **Benefit:** In the massive amounts of data that financial firms handle, data engineering aids in the detection of oddities. Data engineering uses this capacity to protect businesses' and their clients' interests.

- **Example Use Case:** In order to effectively detect abnormalities and prevent fraud, PayPal has deployed data engineering and machine learning algorithms. To minimize fraudulent activity, they examine consumer spending trends and create predetermined patterns.

Customer Relationship Management or CRM

- **Use case:** Customer information is sensitive and very personal. Identity theft and other problems with income loss might result from exposure to sensitive personal data. However, businesses may offer individualized and customer-focused services only they can use this data properly. By facilitating the integration of all client data from several sources, such as transaction histories, selected preferences, and more, data engineering aids in this endeavour. This information is kept in a centralised system that safeguards private information and offers a "eagle eye" into consumer trends and behaviour.

- **Benefit:** To establish a comprehensive client profile, financial institutions meticulously examine and separate consumer data. Institutions use data engineering to create customized products that meet the demands of their clients, giving them a competitive edge.

- **Example Use Case:** One of the most well-known international financial firms, American Express, combines client data and behaviour using data engineering techniques. This makes it easier to create unique financial offers, thrilling incentives, loyalty plans, and more tailored to the requirements of every client (Kim, 2000)

c. Ecommerce

A number of essential elements make up data engineering, which combines to provide a strong basis for data-driven decision making.:

1. **Data Collection:** E-commerce businesses must record a variety of data, such as sales transactions, website interactions, and customer information. Accurate and effective collection of the required data is ensured by data engineering.

2. **Data Processing:** The raw data must be processed and converted into a format that can be used once it has been gathered. To make analysis and decision-making easier, data engineering makes it possible to extract, manipulate, and load data.

3. **Data Storage:** Every day, e-commerce businesses deal with massive amounts of data. To store and manage the constantly expanding amount of data, data engineering aids in the design and implementation of effective data storage systems, such as data lakes and warehouses.

4. **Data Integration:** Data from e-commerce businesses is frequently dispersed across several platforms and sources. Integrating and combining data from several sources to produce a cohesive business picture is known as data engineering.

5. **Data Quality Assurance:** Data from e-commerce businesses is frequently dispersed across several platforms and sources. Integrating and combining data from several sources to produce a cohesive business picture is known as data engineering.

6. **Data Analytics:** Applying analytical methods to find trends, patterns, and insights in the data is made possible by data engineering. These revelations can guide well-informed company plans and judgements.

7. **Data Governance:** Data governance procedures are included in data engineering in addition to the technological elements. For data management, privacy, security, and compliance, this entails creating rules, processes, and guidelines.

8. **Data Visualization:** Following data processing and analysis, data engineering helps to clearly and meaningfully visualise the results. Stakeholders may better comprehend complicated data sets and make wise decisions by using data visualisation techniques.

9. **Data Scalability:** Data engineering is crucial to ensure that systems and infrastructure can expand to meet the growing needs as data volumes continue to rise rapidly. This entails creating and putting into use scalable technologies and structures.

Data engineering helps e-commerce businesses to fully utilize their data and promote data-driven decision-making at all organisational levels by concentrating on five essential elements (Bean, 2003).

7.6 Collaborative Approaches and Best Practices in Data Engineering

Coordination of data engineering practices and processes alongside other organizations are crucial in enhancing effectiveness and efficiency of data management, enhancing productivity and smooth interaction between multiple platforms within organizations. Cross-functional data engineering is where the data engineers, data scientists and analysts, IT specialists create, maintain and optimize the data pipeline. This collaboration improves the exchange of information and ideas, which helps in effective coordination on how data processes meet the requirements of various users, and solves data quality, availability, and growth issues.

Another is on the choice of architectural style that forms one of the best practices in data engineering. This helps teams to be able to create dynamic data management frameworks that address scaling data volume and data variety. This is possible with help of technologies, such as cloud-based data warehouses (Amazon Redshift, Google Big Query) and distributed computing frameworks (Apache Spark), which allow for providing necessary scalable infrastructure for an organization's data needs. It also makes it possible to have the various systems modular to avoid disruptions when updating and maintaining the pipeline.

The second-best practice to be discussed is concentrating a significant effort in achieving high quality of data along with data quality consistency across the data life cycle. Automated data validation, cleansing and transforming are also established to minimize errors and inconsistency of data. Both the data governance framework, which defines possession of data, access rights, and records of access, must be in place to safeguard and legal requirements. Checking for data frequently and also performing their monitoring is very vital to ensure that the required data pipeline is not producing erroneous results due to lack of proper version control procedures.

In addition, DevOps practices are employed in data engineering to enable better development, testing, and deployment. CI/CD methodologies enable the enhancement of an automated process of testing and software release of data pipeline updates, therefore improving efficiency. It can also be useful to containerize application environments using tools like Docker to make it easy for multiple contributors and distributed teams to work on environments that will run on multiple platforms.

Another important best practice is documentation of the team's work. Documentation about data architectures, pipeline work flows and any changes done to the data helps to make everyone understand well and, in the future, if there is a change of guard or if the project is to be handed over to other people, everything will be in order. Documentation also enhances transparency; as opposed to going around in circles trying to identify errors, it tightens up time spent identifying and rectifying bugs inherent in large data networks.

Last but not least, the ability to develop members' education within the team and enhanced knowledge sharing is crucial for anticipating the best new trends and technologies in data engineering. The team is informed about new technologies and how to use them properly to preserve the effectiveness, security, and elasticity of data engineering systems that will be constructed in the future because data processes are changing dynamically as a result of advancements in AI, machine learning, and automation.

7.7 Emerging Opportunities in Data Engineering Careers

Data engineering offers its workers a wide range of job options and contributes to obesity in a rapidly evolving field. The list of common career routes in data science and data engineering fields is provided below:

1. **Data Engineer/Analyst:** Data engineers and analysts are essential to the design, implementation, and upkeep of a data pipeline, which receives, stores, processes, and analyses data. To guarantee data availability and accuracy, they collaborate with data scientists and analysts. Data engineers are in charge of designing and validating data models, data pipelines, and maintaining the integrity and quality of data. For efficient and successful data processing and analysis, the data engineers work directly in a professional partnership with data scientists and analysts.

2. **Big Data Engineer:** Big data engineers are highly skilled in methods for processing and storing enormous amounts of data. They build solutions to expand hardware and software for data storage and analysis, leverage distributed computing architectures, and use sophisticated data processing and systems. Large-scale data processing and storage systems can be managed by big data engineers, who specialised in big data technologies, including Hadoop, Spark, and NoSQL databases. They take part in creating and using technologies for data analysis and processing vast volumes of information.

3. **Machine Learning Engineer:** The development and use of machine learning models to the stability of production systems is the responsibility of machine learning engineers. To integrate data pipelines, implement models, and improve model performance, they collaborate with data scientists, engineers, and software applications. Machine learning engineers collaborate with data engineers to achieve the creation of machine learning models, even if they are not explicitly data engineers. Thus, their deep grasp of scalability and data architecture is one of their greatest advantages.

4. **Data Architect:** Through data management, the data architects provide answers to the organization's issues that help the business stakeholders as much as possible. In addition to monitoring database development and data optimisation, they create data models and establish data governance standards. Data architects are the craftspeople who work on creating the structural and global architecture of data systems. They create data governance rules, model data structures, and build data storage to ensure data availability and consistency.

5. **Data Science Engineer:** Data science engineers may create scalable data pipelines and implement machine learning models to bridge the gap between the two academic specialties, data science faculty and data engineering faculty. They use data systems engineering and programming to put their skills to use in data-driven decision making. ETL (Extract, Transform, Load) developers are experts that create and configure ETL (Extract, Transform, Load) procedures to extract data from several sources and transform it into a format that can be added to databases or data warehouses of choice.

6. **Data Infrastructure Engineer:** Building and maintaining robust data infrastructure, which is made up of databases, data warehouses, and data lakes, is the focus of data infrastructure engineers. They support a variety of data-intensive applications by offering scalability, dependability, and data findability. Deploying and managing data applications on the cloud, which integrates several technologies like AWS, Azure, and Google Cloud, is a new skill brought about by the deeper integration of cloud computing. They store and analyze data in the cloud.

Qualifications for a career in Data Engineering

- **Bachelor's Degree:** Participants in computer science, information technology, engineering, mathematics, or any other related field are typically the target audience for events.

- **Technical Certifications:** Data engineering certification, cloud computing platforms (like AWS Certified Data Engineer), and particular tools and technologies like Apache Spark.
- **Advanced Degrees:** A master's or doctoral degree in data science, data engineering, computer science, or a related field might be the sole prerequisite for a senior role or a specialised one.
- **Experience:** Practical experience with engines, databases, big data technologies, and cloud platforms through educational programs such as internships, projects, or professional employment.
- **Analytical Skills:** Here, critical thinking is crucial, involving data analysis, pattern recognition, and subsequent conclusion and inference making.
- **Communication Skills:** Technical skills like elucidating intricate scientific ideas are important, but so are teamwork, project collaboration, and presenting findings to stakeholders and other audiences.
- **Continuous Learning:** Desire to participate in such professional development programs, stay up to date on industry trends, and stay familiar with technology (Halwani et al., 2022).

7.8 Chapter Conclusion

the chapter re-emphasizes the practical orientation of data engineering applications and their potential for change in various industries. Flexible systems and near real-time processing provide ways through which organizations can address emerging business requirements, facilitate better decision making and optimize existing processes. Despite these considerations that speak to the difficulty of large-scale data implementation, it is possible to overcome them by integrating superior approaches that are in harmony with the contemporary technologies. Also, with the increasing need for insights on people's choices and behaviors, there is a large market for those with knowledge in data engineering. Through knowledge of current trends and development in data engineering, organizations and persons can fully unlock ways to achieve growth, sustainability and prosperity whether on the short term or in the long run within a world that is rapidly becoming data-driven.

Multiple Choice Questions (MCQs)

1. Which company is known for its microservices-based architecture for real-time analytics?

- a. Google
- b. Netflix
- c. IBM
- d. Oracle

2. What is a key application of real-time analytics in finance?

- a. Batch reporting
- b. Fraud detection
- c. Data archiving
- d. Manual audits

3. What is the benefit of using distributed computing frameworks like Apache Spark?

- a. Reducing cloud storage
- b. Enabling parallel data processing
- c. Limiting data scalability
- d. Increasing manual interventions

4. How does predictive maintenance benefit manufacturing?

- a. By delaying system updates
- b. By preventing equipment failures
- c. By limiting real-time data processing
- d. By increasing system downtime

5. Which cloud provider is known for its scalable data platform services?

    a. AWS

    b. Excel

    c. PostgreSQL

    d. Apache Flink

6. What is the main challenge of handling semi-structured data?

    a. Limited scalability

    b. Complex data integration

    c. Real-time data transmission

    d. Simplified data governance

7. What is one key insight from industry applications of data engineering?

    a. Automation reduces workflow efficiency

    b. Scalable platforms improve decision-making

    c. Data democratization is unnecessary

    d. Cloud solutions limit data access

8. What is the role of workflow orchestration tools in data pipelines?

    a. Encrypting data

    b. Automating and managing workflows

    c. Reducing cloud storage costs

    d. Limiting data flow

9. Which tool is commonly used for real-time data processing?

    a. Microsoft PowerPoint

    b. Apache Flink

    c. Excel

    d. Tableau

10. How do scalable platforms impact real-time decision-making?

a. By delaying insights
b. By enabling quick responses to ongoing events
c. By reducing data availability
d. By complicating data processing

Answer

1	2	3	4	5	6	7	8	9	10
b	b	b	b	a	b	b	b	b	b

BIBLIOGRAPHY

Abadi, D. (2012). Consistency Tradeoffs in Modern Distributed Database System Design: CAP is Only Part of the Story. *Computer*. https://doi.org/10.1109/mc.2012.33

Achanta, A., & Bo, R. (2023). Evolving Paradigms of Data Engineering in the Modern Era: Challenges, Innovations, and Strategies. *International Journal of Science and Research (IJSR)*, *12*, 606–610. https://doi.org/10.21275/SR231007071729

ACTIAN. (2025). *How To Build Scalable Data Architectures*. ACTIAN.

Airbyte. (2024). *Real-Time Data Processing: Architecture, Tools & Examples*. Airbyte.

Akermi, M., Hadj Taieb, M. A., & Ben Aouicha, M. (2023). Data Virtualization Enabling Distributed Data Architectures: Data Fabric and Data Mesh. *International Journal of Computer Information Systems and Industrial Management Applications*.

Antova, L., Baldwin, R., Gu, Z., & Waas, F. M. (2019). An Integrated Architecture for Real-Time and Historical Analytics in Financial Services. *Lecture Notes in Business Information Processing*. https://doi.org/10.1007/978-3-030-24124-7_3

Armbrust, M., Fox, A., Griffith, R., Joseph, A. D., Katz, R., Konwinski, A., Lee, G., Patterson, D., Rabkin, A., Stoica, I., & Zaharia, M. (2010). A view of cloud computing. In *Communications of the ACM*. https://doi.org/10.1145/1721654.1721672

Baik, J. (Sophia). (2020). Data privacy against innovation or against discrimination?: The case of the California Consumer Privacy Act (CCPA). *Telematics and Informatics*. https://doi.org/10.1016/j.tele.2020.101431

BasuMallick, C. (2022). *What Is Data Modeling? Process, Tools, and Best Practices*. SpiceWorks.

Battiston, I. (2023). Improving Data Minimization through Decentralized Data Architectures. *CEUR Workshop Proceedings*.

Bean, J. (2003). Engineering global e-commerce sites: A guide to data capture, content, and transactions. In *Elsevier Science*.

Ben Lutkevich. (2023). *What is Data Ingestion_ - Definition from WhatIs*.

Bhathal, G. S., & Singh, A. (2020). *Big Data Computing with Distributed* (Issue June). Springer Singapore. https://doi.org/10.1007/978-981-13-3765-9

Braik, W., Morandat, F., Falleri, J. R., & Blanc, X. (2016). Real time streaming pattern detection for ecommerce. *Proceedings of the ACM Symposium on Applied Computing*. https://doi.org/10.1145/2851613.2851653

Bussa, S. (2024). Evolution of Data Engineering in Modern Software Development. *Journal of Sustainable Solutions*, *1*, 116–130. https://doi.org/10.36676/j.sust.sol.v1.i4.43

Buys, M. (2017). Protecting personal information: Implications of the protection of personal information (POPI) act for healthcare professionals. In *South African Medical Journal*. https://doi.org/10.7196/SAMJ.2017.v107i11.12542

Cao, L. (2010). Domain-driven data mining: Challenges and prospects. *IEEE Transactions on Knowledge and Data Engineering*. https://doi.org/10.1109/TKDE.2010.32

Carvalho, G., Mykolyshyn, S., Cabral, B., Bernardino, J., & Pereira, V. (2022). Comparative Analysis of Data Modeling Design Tools. *IEEE Access*. https://doi.org/10.1109/ACCESS.2021.3139071

Chockalingam, S., & Maathuis, C. (2022). Ontology for Effective Security Incident Management. *International Conference on Cyber Warfare and Security*. https://doi.org/10.34190/iccws.17.1.6

Contribution, C. (2024). *The Role of Stream Processing in Real-Time Analytics - RisingWave_ Open-Source Streaming Database*.

Corodescu, A. A., Nikolov, N., Khan, A. Q., Soylu, A., Matskin, M., Payberah, A. H., & Roman, D. (2021). Big data workflows: Locality-aware orchestration using software containers. *Sensors.* https://doi.org/10.3390/s21248212

Dasppatnaik, S. (2024). *(6) The Modern Data Ecosystem.*

De Lucca, N., Martins, G. M., & Queiroz, R. C. Z. (2023). Brazilian General Data Protection Law (LGPD) and California Consumer Privacy Act (CCPA). *Brazilian Journal of Law, Technology and Innovation.* https://doi.org/10.59224/bjlti.v1i1.38-57

Dearmer, A. (2023). *How to Monitor and Debug Your Data Pipeline.* Integrate.Io.

DeCandia, G., Hastorun, D., Jampani, M., Kakulapati, G., Lakshman, A., Pilchin, A., Sivasubramanian, S., Vosshall, P., & Vogels, W. (2007). Dynamo: Amazon's highly available key-value store. *SOSP'07 - Proceedings of 21st ACM SIGOPS Symposium on Operating Systems Principles.*

Deekshith, A. (2019). Integrating AI and Data Engineering: Building Robust Pipelines for Real-Time Data Analytics. *Double Peer Reviewed.*

DigitalDefynd, T. (2024). *Evolution of Data Engineering [Past, Present & Future] [2025] - DigitalDefynd.*

Dutta, A. (2024). *Data Ingestion: Pipelines, Frameworks, and Process Flows.* FirstEigen.

Erl, T., Cope, R., & Naserpour, A. (2020). Book Review - Cloud Computing Design Patterns. *ACM SIGSOFT Software Engineering Notes.* https://doi.org/10.1145/3385678.3385690

Evans, H. (2023). *Data Engineering Challenges and How to Overcome Them - Velvetech.*

Faggini, M., Cosimato, S., Nota, F. D., & Nota, G. (2019). Pursuing sustainability for healthcare through digital platforms. *Sustainability (Switzerland).* https://doi.org/10.3390/su11010165

frontierinternet. (2022). „The evolution of data storage". In *Преузето: Januaр 2022.* (p. dadas).

GeeksforGeeks. (2024a). *Difference between Batch Processing and Real Time Processing System.* GeeksforGeeks.

GeeksforGeeks. (2024b). *Low latency Design Patterns.* GeeksforGeeks.

Gillis, A. S., Lutkevich, B., & Vaughan, J. (2023). *data engineer*.

Gupta, I., Singh, A. K., Lee, C. N., & Buyya, R. (2022). Secure Data Storage and Sharing Techniques for Data Protection in Cloud Environments: A Systematic Review, Analysis, and Future Directions. In *IEEE Access*. https://doi.org/10.1109/ACCESS.2022.3188110

Halwani, M. A., Amirkiaee, S. Y., Evangelopoulos, N., & Prybutok, V. (2022). Job qualifications study for data science and big data professions. *Information Technology and People*. https://doi.org/10.1108/ITP-04-2020-0201

Hechler, E., Weihrauch, M., & Wu, Y. (2023). Data Fabric and Data Mesh Use Case Scenarios. In *Data Fabric and Data Mesh Approaches with AI*. https://doi.org/10.1007/978-1-4842-9253-2_3

Hellemans, I., Porter, A. J., & Diriker, D. (2022). Harnessing digitalization for sustainable development: Understanding how interactions on sustainability-oriented digital platforms manage tensions and paradoxes. *Business Strategy and the Environment*. https://doi.org/10.1002/bse.2943

Highleyman, B., Holenstein, B., & Holenstein, P. J. (2005). *Fault tolerance vs. High Availability* (Issue October, pp. 9–10).

HitachiSolutions. (2025). *Data Fabric and Data Mesh: Does the Future Need You?* HitachiSolutions.

Huth, M., & Nielson, F. (2019). Static analysis for proactive security. In *Lecture Notes in Computer Science (including subseries Lecture Notes in Artificial Intelligence and Lecture Notes in Bioinformatics)*. https://doi.org/10.1007/978-3-319-91908-9_19

Iabacbdmur. (2023). *Tools of Data Engineering _ the Modern Data Ecosystem - IABAC*.

Ibrahem, A., & Zeebaree, S. (2024). Tackling the Challenges of Distributed Data Management in Cloud Computing -A Review of Approaches and Solutions. *International Journal of Intelligent Systems and Applications in Engineering, 12*, 340–355.

Janev, V. (2021). *Semantic Intelligence in Big Data Applications. December*. https://doi.org/10.1007/978-3-030-76387-9

Jani, Y. (2024). Unified Monitoring for Microservices: Implementing Prometheus and Grafana for Scalable Solutions. *Journal of Artificial Intelligence, Machine*

Learning and Data Science, 2, 848–852. https://doi.org/10.51219/JAIMLD/yash-jani/206

Jayabalan, D. (2024). Enhancing Machine Learning Life Cycle through Advanced Data Engineering. *International Journal of Computer Trends and Technology, 7*(4), 136–139.

Kim, S. H. (2000). An architecture for advanced services in cyberspace through data mining: A framework with case studies in finance and engineering. *Journal of Organizational Computing and Electronic Commerce.* https://doi.org/10.1207/S15327744JOCE1004_04

Koczwara, Ł. (2023). *Data Ecosystem Explained – Key Elements & Benefits of Modern Data Solutions.*

Kosambia, S. (2021). *Designing Modern Data Platform on Cloud.*

Kreps, J., Narkhede, N., & Rao, J. (2011). Kafka: a Distributed Messaging System for Log Processing. *ACM SIGMOD Workshop on Networking Meets Databases.*

Krishna, V. (2023). *The Evolution of Data Engineering_ A Walk Through Time _ by vamshi krishna _ Medium.*

Kumar, K. (2021). Integrated benchmarking standard and decision support system for structured, semi structured, unstructured retail data. *Wireless Networks.* https://doi.org/10.1007/s11276-021-02843-4

Kutay, J. (2024). *Real-Time Analytics Use Cases and Examples.* Striim.

Lakshman, A., & Malik, P. (2010). Cassandra - A decentralized structured storage system. *Operating Systems Review (ACM).* https://doi.org/10.1145/1773912.1773922

Lechler, S., Canzaniello, A., Roßmann, B., von der Gracht, H. A., & Hartmann, E. (2019). Real-time data processing in supply chain management: revealing the uncertainty dilemma. *International Journal of Physical Distribution and Logistics Management.* https://doi.org/10.1108/IJPDLM-12-2017-0398

Martin, Y. S., & Kung, A. (2018). Methods and Tools for GDPR Compliance Through Privacy and Data Protection Engineering. *Proceedings - 3rd IEEE European Symposium on Security and Privacy Workshops, EURO S and PW 2018.* https://doi.org/10.1109/EuroSPW.2018.00021

Miser, E., & Sarioguz, O. (2024). Data-Driven Decision-Making: Revolutionizing Management in the Information Era. *International Research*

Journal of Modernization in Engineering Technology and Science. https://doi.org/10.56726/IRJMETS49577

Moore, W., & Frye, S. (2019). Review of HIPAA, Part 1: History, protected health information, and privacy and security rules. *Journal of Nuclear Medicine Technology.* https://doi.org/10.2967/JNMT.119.227819

Mullins, C. S. (2022). *What is ELT_ How is it Different from ETL_ _ TechTarget.*

Mumuni, A., & Mumuni, F. (2024). Automated data processing and feature engineering for deep learning and big data applications: A survey. *Journal of Information and Intelligence.* https://doi.org/10.1016/j.jiixd.2024.01.002

Mutiso, J. (2024). *(6) Essential System Design Principles for Scalable Architectures and The Role of Fault Tolerance in Modern System Design.*

Nineleaps. (2021). *The Evolution of Data Engineering _ by Nineleaps _ Technology at Nineleaps _ Medium.*

NINELIPES. (2023). *Understanding the Data Engineering Evolution.*

Office of the Privacy Commissioner of Canada. (2019). *PIPEDA fair information principles.* Government of Canada.

Orangemantra. (2022). *AWS vs Azure vs GCP_ Cloud Platform to Choose for Business_.*

Pappas, I. O., Mikalef, P., Giannakos, M. N., Krogstie, J., & Lekakos, G. (2018). Big data and business analytics ecosystems: paving the way towards digital transformation and sustainable societies. In *Information Systems and e-Business Management.* https://doi.org/10.1007/s10257-018-0377-z

Parida, P. R. (2024). Integrating IoT with AI-Driven Real-Time Analytics for Enhanced Supply Chain Management in Manufacturing. *Journal of Artificial Intelligence Research and Applications, 4*(2), 40–84.

Peruccon, A., & Simeone, L. (2023). Designing Systems in a Fine-Grained and Inclusive Way, by Integrating Service Design within Futures Studies and Foresight. *Design Management Journal.* https://doi.org/10.1111/dmj.12084

Pogiatzis, A., & Samakovitis, G. (2021). An event-driven serverless etl pipeline on aws. *Applied Sciences (Switzerland).* https://doi.org/10.3390/app11010191

Ponnusamy, S., & Gupta, P. (2023). Connecting the Dots: How Data Lineage Helps in Effective Data Governance. *International Journal of Computer Science and Engineering.* https://doi.org/10.14445/23488387/ijcse-v10i10p102

Pratt, L. (2024). *4 common data engineering challenges you need to overcome to succeed _ by Lynne Pratt _ Medium.*

QLICK. (2023). *Data Lake vs Data Warehouse.*

Raschka, S., Patterson, J., & Nolet, C. (2020). Machine learning in python: Main developments and technology trends in data science, machine learning, and artificial intelligence. In *Information (Switzerland).* https://doi.org/10.3390/info11040193

Richman, J. (2024a). *What Is An ETL Pipeline? Examples & Tools (Guide 2024).* Estuary.

Richman, J. (2024b). *What Is Real-Time Processing (In-depth Guide For Beginners).* Estuary.

Sadeghi, A. (2024). *Open Source Data Engineering Landscape 2024 _ by Alireza Sadeghi _ Medium.*

Santos, S., Gonçalves, H. M., & Teles, M. (2023). Social media engagement and real-time marketing: Using net-effects and set-theoretic approaches to understand audience and content-related effects. *Psychology and Marketing.* https://doi.org/10.1002/mar.21756

Sarker, I. H. (2021). Machine Learning: Algorithms, Real-World Applications and Research Directions. *SN Computer Science.* https://doi.org/10.1007/s42979-021-00592-x

Science, D., & Engineering. (2023). Data Engineering: Data Warehouse, Data Pipeline and Data Engineer Role | AltexSoft. In *Blog Altexsoft.*

Shahrivari, S. (2014). Beyond batch processing: Towards real-time and streaming big data. In *Computers.* https://doi.org/10.3390/computers3040117

Sheldon, R. (2021). *How to choose between SQL and NoSQL databases.*

Sherman, R. (2015). *Data Integration Team - an overview _ ScienceDirect Topics.*

Silva, B. N., Khan, M., Jung, C., Seo, J., Muhammad, D., Han, J., Yoon, Y., & Han, K. (2018). Urban planning and smart city decision management empowered by real-time data processing using big data analytics. *Sensors (Switzerland).* https://doi.org/10.3390/s18092994

Simplilearn. (2024). *Top Data Engineering Tools 2024_ Unleash Your Potential.*

Singh, I., & Singh, B. (2023). Access management of IoT devices using access control mechanism and decentralized authentication: A review. *Measurement: Sensors.* https://doi.org/10.1016/j.measen.2022.100591

Smartbear. (2022). *Introduction to Performance Monitoring Metrics | SmartBear.*

Sun, L., Zhang, H., & Fang, C. (2021). Data security governance in the era of big data: status, challenges, and prospects. In *Data Science and Management.* https://doi.org/10.1016/j.dsm.2021.06.001

Sun, X., He, Y., Wu, D., & Huang, J. Z. (2023). *Survey of Distributed Computing Frameworks for Supporting Big Data Analysis.*

SustainableDataPlatform. (2024). *Data Tool House for evidence-based climate action.* SustainableDataPlatform.

Syn-Hershko, I. (2022). *Architectures of a Modern Data Platform.*

Truong, N., Sun, K., Wang, S., Guitton, F., & Guo, Y. K. (2021). Privacy preservation in federated learning: An insightful survey from the GDPR perspective. *Computers and Security.* https://doi.org/10.1016/j.cose.2021.102402

Tucker, D. (2025). *Data Mesh or Data Fabric? Do Better with Both.* Booz\ Allen\ Hamilton.

Von Dietze, A., & Allgrove, A. M. (2014). Australian privacy reforms-an overhauled data protection regime for Australia. *International Data Privacy Law.* https://doi.org/10.1093/idpl/ipu016

Wang, L., & Alexander, C. A. (2020). Big data analytics in medical engineering and healthcare: methods, advances and challenges. In *Journal of Medical Engineering and Technology.* https://doi.org/10.1080/03091902.2020.1769758

Wickramasinghe, S. (2024). *AWS vs Azure vs GCP_ Comparing The Big 3 Cloud Platforms.*

William McGeveran. (2019). The Duty of Data Security. *Minnesota Law Review.*

Wills, M. J. (2014). Decisions through data: Analytics in healthcare. *Journal of Healthcare Management.* https://doi.org/10.1097/00115514-201407000-00005

Yadranjiaghdam, B., Pool, N., & Tabrizi, N. (2017). A survey on real-time big data analytics: Applications and tools. *Proceedings - 2016 International Conference on Computational Science and Computational Intelligence, CSCI 2016.* https://doi.org/10.1109/CSCI.2016.0083

ABOUT THE AUTHOR

Amarnath Immadisetty is a seasoned technology leader with over 18 years of experience in software engineering and a deep passion for leveraging cutting-edge technologies to drive innovation. At the time of writing this book, he is serving as the Senior Manager of Software Engineering at Lowe's, Amarnath leads a team of more than 20 engineers, guiding them to deliver impactful solutions in customer data platforms, software development, and large-scale data analytics. His work has significantly enhanced business performance by improving data-driven decision-making and operational efficiency.

Throughout his career, Amarnath has held key roles at notable companies such as Lowe's, Target, Uniqlo, and CMC Limited. His expertise spans a wide array of areas, including software engineering, large-scale system architectures, and advanced data analytics. Amarnath's strong technical leadership and ability to drive transformation through innovative solutions have made him a respected figure in the tech industry.

In addition to his technical accomplishments, Amarnath is passionate about mentoring the next generation of engineers and actively advocates for diversity and inclusion in tech. He firmly believes that diverse teams foster better innovation and problem-solving.

www.ingramcontent.com/pod-product-compliance
Lightning Source LLC
Chambersburg PA
CBHW040726120726
48010CB00001B/20